Administrating Microsoft Dynamics 365 Business Central Online

A practical guide to SaaS administration and migration from your on-premise Business Central environments to the cloud

Andrey Baludin

BIRMINGHAM—MUMBAI

Administrating Microsoft Dynamics 365 Business Central Online

Copyright © 2022 Packt Publishing

Group Product Manager: Alok Dhuri
Publishing Product Manager: Harshal Gundetty
Senior Editor: Ruvika Rao
Technical Editor: Pradeep Sahu
Copy Editor: Safis Editing
Project Coordinator: Deeksha Thakkar
Proofreader: Safis Editing
Indexer: Hemangini Bari
Production Designer: Jyoti Chauhan
Marketing Coordinators: Deepak Kumar and Rayyan Khan
Business Development Executive: Uzma Sheerin

First published: July 2022

Production reference: 1280622

Published by Packt Publishing Ltd.
Livery Place
35 Livery Street
Birmingham
B3 2PB, UK.

ISBN 978-1-80323-480-9

www.packt.com

To the memory of my grandparents Ivan, Maria, and Lidia. To my wife, Anastasia, and daughter, Daria, for being the meaning of my life. To my parents, Svetlana and Igor, and to my grandpa Leonid for your faith in me.

– Andrey Baludin

Contributors

About the author

Andrey Baludin is a solution architect at Awara IT Solutions (a Microsoft Gold partner). His educational qualifications include a master's degree from St. Petersburg Polytechnic University. Andrey is a Microsoft MVP. He has been working with Microsoft Dynamics NAV since 2008, from 5.0 to the latest versions of Dynamics 365 Business Central. Andrey is a real fan of Business Central and Azure, and his specialization is integrating Business Central with everything.

I want to thank my employer, Awara IT, for projects, which made this book real, and the brilliant Packt team, who supported me throughout the process.

About the reviewers

Stefano Demiliani is a Microsoft MVP, MCT, and CTO, an Azure solution architect, and a long-time expert on different Microsoft technologies. He's a trainer for Microsoft IT and WE and a speaker at many international conferences. He has worked with Packt Publishing on many IT books, and he's the author of some of the most successful books in the Microsoft Dynamics world. You can reach him on Twitter (@demiliani), LinkedIn, or his site (www.demiliani.com).

Steven Renders is a trainer/consultant, with skills spanning business and technical domains, with more than 20 years of experience. He provides training and consultancy focused on Microsoft Dynamics 365 Business Central and Power BI. Steven is a partner at Plataan.tv, based in Belgium. Plataan is a leading offline and online learning company. It helps organizations and individuals excel through learning and training programs.

On January 1, 2016, Steven received the Microsoft MVP Award. This award is given to exceptional technical community leaders who actively share their high-quality, real-world expertise with others. Steven has authored two books, namely *Microsoft Dynamics NAV 2009 – Professional Reporting* and *Microsoft Dynamics NAV 2015 – Professional Reporting*. He specializes in Microsoft Dynamics 365 Business Central and business intelligence and reporting.

Table of Contents

3

Environment Details and Notifications Setup

4

Telemetry Setup and Analysis

5

Reported Outages and Operations

6

Tenant Capacity Management

7

Admin Center APIs

Part 2: Dynamics 365 Business Central Cloud Migration Tool

8

Cloud Migration Schema and Limitations

9

Cloud Migration Setup

10

Migration Process

11

The Real Migration Experience

Index

Other Books You May Enjoy

Preface

This book keeps all the needed information in one place. It contains detailed information about the Admin portal and cloud migration process extended with a real usage experience. The book will guide you from basic operations to advanced automation processes, from a few clicks in web interface to scripts writing.

Who this book is for

The first part of the book will help all Dynamics 365 Business Central users to be deeply integrated into the administration processes. Beginners will learn the basic things; experienced users will be interested in a telemetry setup and analysis and processes automation. The second part of this book mainly caters to Dynamics 365 Business Central developers and consultants who are planning to move their on-premises solution to the cloud.

What this book covers

Chapter 1, *Overview of the Dynamics 365 Business Central Admin Center*, contains common information about how to start your work with Dynamics 365 Business Central Admin Center and get access if you cannot open it. In addition, you will get details about environment types, limits, and planning of your free space.

Chapter 2, *Managing Business Central Environments*, is about managing your Business Central environments, their types, statuses, updates information, and the creation of new environments. In addition, it contains information about how to create a copy of your production environment and test your apps with future releases.

Chapter 3, *Environment Details and Notification Setup*, contains detailed information about your environment, versions management, apps and sessions, database management, backup and restore.

Chapter 4, *Telemetry Setup and Analysis*, explains how production environment debugging is quite tricky when your Business Central is SaaS. Telemetry will help you with that. In this chapter, you will learn how to check basic telemetry on the Admin portal and how to set up extended telemetry with App Insights.

Chapter 5, Reported Outages and Operations, helps you to get details about reported outages and monitor your app's operations. Here, you will learn how to report an outage, search your reported outages, and list your installed apps without opening your Business Central environments.

Chapter 6, Tenant Capacity Management, looks at how SaaS environments have some capacity limitations. Learn how to get details about your database's capacity, environment limits, most massive data entities, and how to keep your data within limits. In addition, we will learn what Microsoft determines as a "Big Customer" for which it could be better to use an on-premises environment.

Chapter 7, Admin Center API, explores ways you could perform almost all actions we listed before without opening the Admin Center. We can use the Admin Center API for that. You can automate routine tasks with scripts. Create and delete environments, schedule updates, and set up notifications – all these operations could be run with a few clicks by calling the APIs.

Chapter 8, Cloud Migration Schema and Limitations, teaches you when you would want to use cloud migration, functions of the cloud migration tool, ways of usage, work schema, and main nodes.

Chapter 9, Cloud Migration Setup, provides step-by-step instructions that help you to set up your cloud migration.

Chapter 10, Migration Process, teaches you what to do when you successfully complete your migration setup. Learn how to run the migration process in two steps, how to check migration results by tables, and what to do when your migration fails.

Chapter 11, The Real Migration Experience, explains how, with our customer, we decided to use cloud migration for data replication just when it appeared and strict limitations were removed. Here is our project's story, which could be useful for you.

To get the most out of this book

For some activities, you need paid Dynamics 365 Business Central and Azure subscriptions.

Software/hardware covered in the book	Operating system requirements
Dynamics 365 Business Central	Windows, macOS
Microsoft self-hosted integration runtime	
PowerShell	
Postman	

Download the color images

We also provide a PDF file that has color images of the screenshots and diagrams used in this book. You can download it here: https://packt.link/GUFuQ.

Conventions used

There are a number of text conventions used throughout this book.

Code in text: Indicates code words in text, database table names, folder names, filenames, file extensions, pathnames, dummy URLs, user input, and Twitter handles. Here is an example: "After you press **Sign In**, you should see that your script executed without errors and access token keeps in the $Token variable."

A block of code is set as follows:

```
$response = Invoke-WebRequest `
    -Method Get `
    -Uri    https://api.businesscentral.dynamics.com/admin/
v2.11/applications/environments `
    -Headers @{Authorization=("Bearer $Token")}
Write-Host (ConvertTo-Json (ConvertFrom-Json $response.
Content))
```

Bold: Indicates a new term, an important word, or words that you see onscreen. For instance, words in menus or dialog boxes appear in **bold**. Here is an example: "Now, we need to provide permissions to your app registration. Click on the **API permissions** tab and then select **+ Add a permission**."

> **Tips or Important Notes**
> Appear like this.

Get in touch

Feedback from our readers is always welcome.

General feedback: If you have questions about any aspect of this book, email us at customercare@packtpub.com and mention the book title in the subject of your message.

Errata: Although we have taken every care to ensure the accuracy of our content, mistakes do happen. If you have found a mistake in this book, we would be grateful if you would report this to us. Please visit www.packtpub.com/support/errata and fill in the form.

Piracy: If you come across any illegal copies of our works in any form on the internet, we would be grateful if you would provide us with the location address or website name. Please contact us at copyright@packt.com with a link to the material.

If you are interested in becoming an author: If there is a topic that you have expertise in and you are interested in either writing or contributing to a book, please visit authors.packtpub.com.

Share Your Thoughts

Once you've read *Administrating Microsoft Dynamics 365 Business Central Online*, we'd love to hear your thoughts! Scan the QR code below to go straight to the Amazon review page for this book and share your feedback.

https://packt.link/r/1803234806

Your review is important to us and the tech community and will help us make sure we're delivering excellent quality content.

Part 1: Dynamics 365 Business Central Admin Center

The Dynamics 365 Business Central admin center enables administrators to manage SaaS environments and everything related to them. In this part, you'll learn how to check your Business Central SaaS environment details, create new environments of different types, set up telemetry and notifications, automate tasks, and keep your cloud data within limits.

This part contains the following chapters:

- *Chapter 1, Overview of the Dynamics 365 Business Central Admin Center*
- *Chapter 2, Managing Business Central Environments*
- *Chapter 3, Environment Details and Notification Setup*
- *Chapter 4, Telemetry Setup and Analysis*
- *Chapter 5, Reported Outages and Operations*
- *Chapter 6, Tenant Capacity Management*
- *Chapter 7, Admin Center API*

1
Overview of the Dynamics 365 Business Central Administration Center

Hey everyone! Many of you know how to manage on-premises Dynamics NAV and Business Central environments with **Server Administration Tool**, but how to do it when your environment is somewhere in the cloud? Do you need to ask your Microsoft Partner, or even Microsoft itself, to create a new sandbox, restore a backup, or restart the environment?

Meet the **Admin Center**, a tool to rule all of your Dynamics 365 Business Central SaaS environments.

This chapter contains common information on how to start your work with the Dynamics 365 Business Central Admin Center. You will learn how to get access to the center in different ways and assign Business Central licenses. In addition, you will get details about environment types, limits, and the planning of your free space.

In this chapter, we are going to cover the following main topics:

- The Admin Center main page

- Required roles

- Environment types and limits

By the end of this chapter, you will be able to set up the required permissions and get access to the Admin Center.

Technical requirements

To have access to the Dynamics 365 Business Central Admin Center, you must be one of the following user types:

- An internal user with a *Global administrator* or *Dynamics 365 administrator* role

- A delegated administrator from the assigned partner with an *Administrator Agent* or *Helpdesk Agent* role

We will learn how to assign these roles in the next sections.

The Admin Center main page

What is the Admin Center? It is not an application you need to install on your laptop. It *lives* directly in your internet browser as a website. You need to open it and sign in with a Microsoft account assigned with the required role.

You can open the Admin Center main page using either of these three ways:

- From the Dynamics 365 Business Central web client

- Using a direct URL

- From the Microsoft Partner Center

Accessing the Admin Center from the web client

1. Open your Dynamics 365 Business Central environment and click on the **Settings** icon in the top-right corner. The **Settings** window will open and you will see the **Admin Center** link:

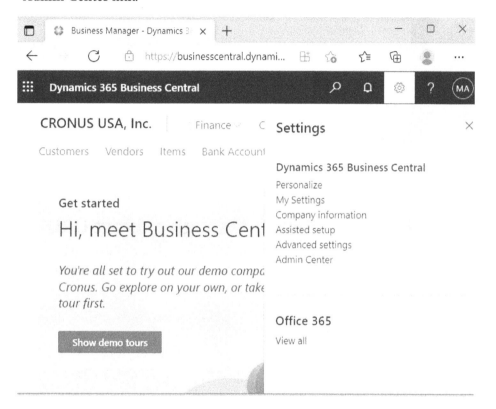

Figure 1.1 – Admin Center in Settings

Important Note

If you cannot see the **Admin Center** link from the web client, this means that you don't have the required permissions described previously.

---

2. Click on the link and you will get the center in a new window:

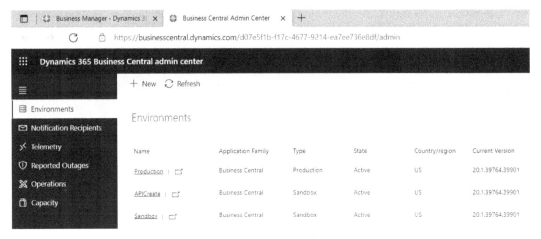

Figure 1.2 – Admin Center home page

Accessing the Admin Center through a direct URL

This is my most-used way of accessing the Admin Center, especially if you have several tenants. You can create and save the Admin Center's URL for each customer's tenant.

Let's look at the center's URL:

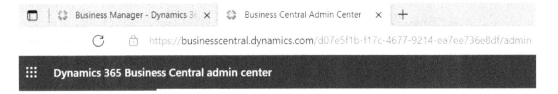

Figure 1.3 – Admin Center URL with the tenant ID

The URL consists of this construction: `https://businesscentral.dynamics.com/<<Tenant ID>>/admin`.

Here, `<<Tenant ID>>` is your tenant ID. So, if you know the ID, you can easily create the Admin Center URL and open it without entering the web client.

The tenant ID is hidden under the **Help & Support** menu. To open it, click on the **Help** icon in the top-right corner and choose the **Help & Support** link:

MA
Help
Search here...
Help & Support
Keyboard Shortcuts

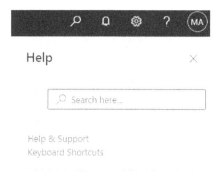

Figure 1.4 – Help menu

At the end of the **Report a problem** section, you will find the tenant ID:

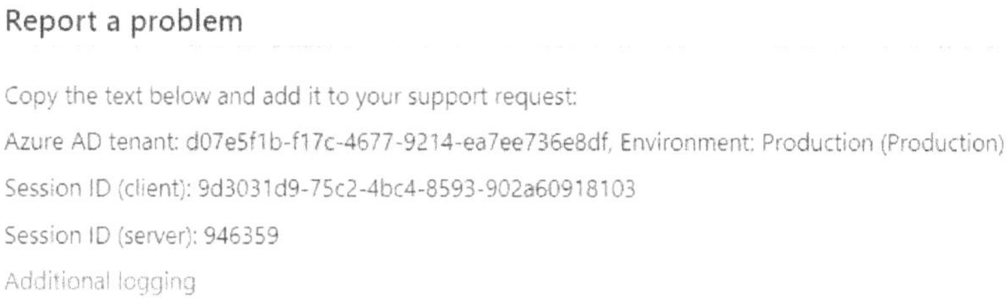

Report a problem
Copy the text below and add it to your support request:
Azure AD tenant: d07e5f1b-f17c-4677-9214-ea7ee736e8df, Environment: Production (Production)
Session ID (client): 9d3031d9-75c2-4bc4-8593-902a60918103
Session ID (server): 946359
Additional logging

Figure 1.5 – Report a problem section

This way, my tenant Admin Center URL will be `https://businesscentral.dynamics.com/d07e5f1b-f17c-4677-9214-ea7ee736e8df/admin`.

However, tenant IDs look weird for the typical user and it's close to impossible to remember it and reproduce it somewhere. Therefore, we can replace the tenant ID with the tenant company's domain and it will work!

If you are an internal user, check your account, for example, `John.Doe@mycompany.com`.

The Admin Center's URL will be `https://businesscentral.dynamics.com/mycompany.com/admin`.

If you are a delegated administrator from an assigned partner, then you need to know your customer's domain to replace the tenant ID. For my demonstration environment, I can access the Admin Center like this:

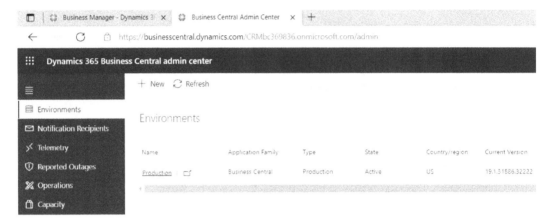

Figure 1.6 – Admin Center URL with domain

Accessing the Admin Center through the Partner Center

If you are a Microsoft Partner who manages customer's environments, then you can access the Admin Center from the Partner Center at `https://partner.microsoft.com/`:

1. Click on **CSP**, then choose **Customers**.
2. Select the relevant customer and click on **Service management**.
3. Then, choose **Dynamics 365 Business Central** under **Administer services**:

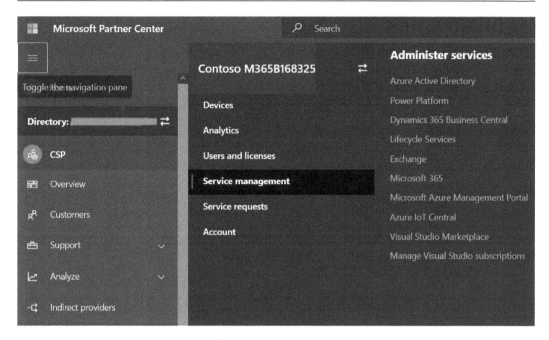

Figure 1.7 – Admin center access from Partner center

This URL will open the customer's Admin Center. This way is useful when you do not know the exact URLs that customers use for their environment.

But, what if you get an **Unauthorized** error (as shown in *Figure 1.8*) after signing in to the Admin Center?

Dynamics 365 Business Central

Unauthorized

Access is allowed only for tenant environment administrators
Make sure the user has Global admin, Helpdesk admin, or Dynamics 365 admin
role assigned.

Figure 1.8 – Unauthorized error

In the next section, we will check which roles you need to assign to your account to get access to the Admin Center.

Required roles

You could be from the Partner or the customer side, I don't know. Not every employee can enter the Admin Center. In this section, we will look at both user types and at the roles that they need to have.

Internal users

For internal users, you must be one of the following to enter:

1. Global administrator
2. Dynamics 365 administrator

If you just need to manage Business Central environments, you can use the **Dynamics 365 administrator** role.

However, in most cases, the person who manages your Business Central infrastructure also creates users and assigns licenses for them, so this way, it is better to have a **Global administrator** role.

In the next screenshot, you can see how the roles assignment looks from the Azure portal side:

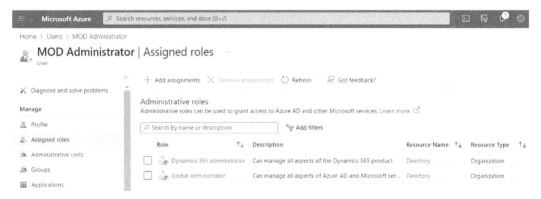

Figure 1.9 – Azure Active Directory roles assignment

In addition, you do not need a Dynamics 365 Business Central license to access the Admin Center or even Business Central itself, but without the license, you will have read-only access.

To assign a license to a user, your Global administrator should complete these steps:

1. Open the **Licenses** page in Azure Active Directory, select **All products** and choose
    your Dynamics 365 Business Central license, Essentials or Premium, depending on
    what type of environment you have. You can see how many licenses you have and
    how many of them are currently in use and available to use:

Figure 1.10 – License list

2. Next, you will see your assigned users list. Then, click **+ Assign**:

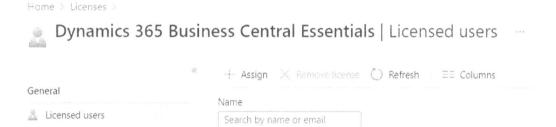

Figure 1.11 – License assignment 1

3. In the final step, click on **+ Add users and groups** to choose the users you want to assign. You can select several users or even user groups:

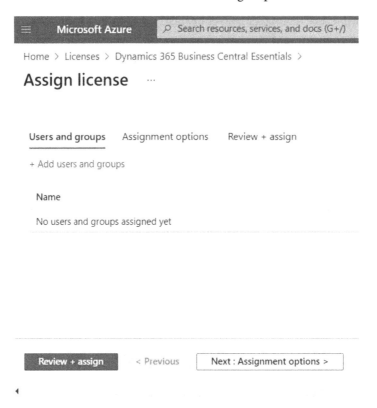

Figure 1.12 – License assignment 2

Delegated admins

A delegated admin is a reselling partner's account that has access to the customer's Dynamics 365 Business Central environment and Admin Center with administrator permissions. This account does not count in Business Central licenses usage, is not visible in the customer's Azure Active Directory, and the customer cannot manage this account, but the customer can remove delegated admin privileges from the partner.

If you are a partner, first of all, the customer must grant you administrator permissions through the relationship request. For this, you need to perform the next actions:

1. Open the **Partner Center**.
2. Open the **CSP | Customers** menu.
3. Click on **Request a reseller relationship**:

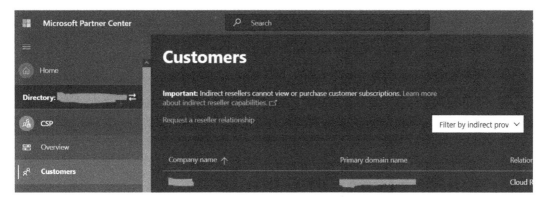

Figure 1.13 – Request a reseller relationship

4. Choose your customer from the list, make sure that the **Include delegated administration privileges for Azure Active Directory and Office 365** box is checked, and send the invitation by clicking the **Done** button:

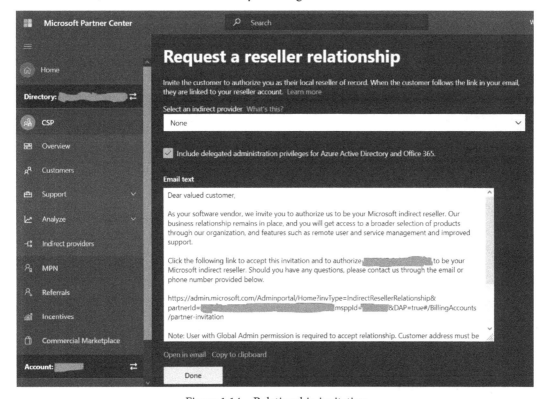

Figure 1.14 – Relationship invitation

After that, you need to be included in one of these two groups:

- **Admin Agents** – These will be Global administrators
- **Helpdesk Agents** – These will be Helpdesk administrators.

Both roles will have access to the Admin Center. The customer could remove this permission from a partner from the Microsoft 365 Admin Center: `https://admin.microsoft.com/AdminPortal/Home#/partners/`.

Environment types and limits

When you work with Dynamics 365 Business Central, you usually need to check some new features and new setup. Developers need to test their code somewhere. For these actions, you need a test environment, a sandbox. In this section, we will look at the production and sandbox environment types and see how many of these environments you can have.

Overview

Let's look at the Admin Center main window:

Figure 1.15 – Admin center overview

Here, you can see your **Environments** list with additional information:

- **Name**
- **Application Family**
- **Type**
- **State**

- **Country/region**
- **Current Version**
- **Available Update Version**
- **Scheduled Update Date**
- **Security Group**

Environments limitations

There are two types of environments:

- Production
- Sandbox

Typically, customers have one production environment; but, for example, if you have companies in different countries with different localization, you might want to have more production environments.

The sandbox environment has a number of differences from the production one. Here are the most important:

- Performance is lower because sandboxes are running on different hardware.
- You cannot export the database for the sandbox environment.
- In the sandbox, you can activate the **Premium user experience** and test the manufacturing and service functionalities.
- You can use the **Designer** page in the sandbox environment.
- You can publish your apps from Visual Studio Code directly to the sandbox, but they will be unpublished after each cumulative update.
- You can debug user sessions in the sandbox.

You can have one production environment and three sandbox environments without additional costs. These are the environments that you can create by yourself. You will see this warning when you try to create a new environment by clicking on the + **New** action:

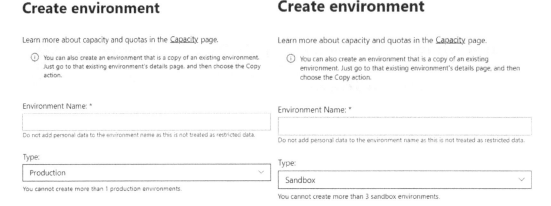

Figure 1.16 – New environment creation

However, you can increase this quota in two ways:

1. Ask your CSP provider to create more production environments and pay a fee. Each production environment will give you three extra sandboxes.

2. Migrate existing environments from other tenants. For example, you have several Azure tenants and decided to merge them. Or your Dynamics 365 Business Central environment from earlier has been created in the wrong tenant and you need to move it to your normal tenant. For this action, you must create a Support request to Microsoft from the Power Platform Admin Center: `https://admin.powerplatform.microsoft.com/account/login/<<Tenant ID>>`.

You can read an explanation of the migration process here: `https://docs.microsoft.com/en-us/dynamics365/business-central/dev-itpro/administration/tenant-admin-center-environments-move`.

Capacity limitations

Until July 2021, we had a strict capacity limitation – 80 GB per tenant, not depending on environment numbers. This was like a stopper for even middle-sized customers. Just imagine that you have a 30 GB production database and you want to create a copy of the production environment to test some new scenarios with production data.

This copy would also take 30 GB of capacity and you would only have 20 GB left. If you need an extra sandbox for development, you cannot create a copy of the production environment without buying a capacity add-on.

For now, we have the next capacity calculation:

Condition	Capacity
Default	80 GB
Each Premium user license	3 GB
Each Essential user license	2 GB
Each device license	1 GB
Each extra production environment	4 GB
Dynamics 365 Business Central Database Capacity Add-on (1 GB)	1 GB
Dynamics 365 Business Central Database Capacity Add-on (100 GB)	100 GB

So, for example, if you have one typical production environment (Essential) with 25 users, you will have an 80-GB default capacity plus a 25 2-GB license capacity, which equals 130 GB of total capacity. That's much more interesting!

> **Important Note**
>
> Each user with a Dynamics 365 Business Central license has access to all tenant environments.
>
> If your environments become out of capacity, this does not mean that all your operations in Business Central will be blocked. You only will not be able to create (or copy) new environments until you get free capacity or buy the capacity add-on.

Summary

You are now able to get access to the Admin Center and if your access is restricted, you know which permissions you require and what actions you need to do. You know how many environments you could create in your tenant and can already plan your future with Business Central SaaS.

In the next chapter, we will get closer to Dynamics 365 environments.

2
Managing Business Central Environments

The **Environments** page is the main page of the Business Central Admin Center, and is the page where you will spend most of your time. It is extremely important in the cloud administration process.

This chapter is about managing your Business Central environments, their types, statuses, updates information, and the creation of new environments. In addition, it contains information about how to create a copy of your production environment and test your apps with future releases.

In this chapter, we are going to cover the following main topics:

- The Environments list
- Creating a production or sandbox environment
- Creating a copy of the existing environment

By the end of this chapter, you will be able to get structured information about your environments, create environments of different types, and rename and delete existing environments.

Understanding the Environments list

In this section, we will have a look at the **Environments** list; this is the first thing that you will see when you open the Admin Center. As the main window, it contains basic and most important information, such as the environment quantity, their names, current states, and upgrade schedules (if possible).

Figure 2.1 – Environments list

On top of the **Environments** list, you can see two extra actions, **+ New** and **Refresh**. **+ New** allows you to create new environments and **Refresh** is used to update environment information on the list. We will take a closer look at these actions in the next section.

The list has the following columns:

- **Name**: This consists of the names of your environments. On my **Environments** list, you can see two environments with the names **Production** and **Sandbox**. These are the default names for your first production and sandbox environments. If you open the Admin Center for the first time, you will find only the production environment there. When you create a new environment, you could type your own name. You could also rename any existing environment. In addition, the **Name** column is a hyperlink to the environment card with additional information and actions about the environment.

Important Note

The environment name cannot be longer than 30 characters and cannot contain spaces or special characters except underscores(_) and dashes (-).

- **Application Family**: This describes about the environment's application family, which is used in the Admin Center's API. The standard value is **Business Central**, but it could have different values for solutions developed by **Independent Software Vendor** (**ISV**) partners.

- **Type**: This can be either **Production** or **Sandbox**. In the **Production** environment, you are doing your work, and in the **Sandbox** environment, you are doing tests and development.

- **State**: The current state of the environment. It usually has the **Active** value, but you could also see the other statuses. They can be as follows:

  I. **Suspended**: This can either mean your Business Central subscription is suspended (for example, you didn't pay for your license) or you didn't use your environment for a long time.

  II. **Not ready**: A short status that appears while you create a new environment or restart an existing one.

  III.**Preparing**: Your environment is either creating, restarting, or upgrading.

  IV. **Active**: Your environment is up and running. You can only use it in this state.

  V. **Removing**: Your environment is being deleted.

- **Country/region**: The localization version of your environment. This could be a country code such as **US**, **FI**, or **GB**, or could have a special value such as **W1**, which means World One, the international version without localization functionalities. The number of available country localizations increases from release to release. The **W1** version environment cannot be created from the Admin Center; you must raise a support ticket with Microsoft for that.

- **Current Version**: This tells you the current version of your environment. For example, `19.1.31886.32222`. You can see four groups of digits:

  - **First digit**: This gives us the major release number. Major releases come twice a year – in April and October as waves 1 and 2, respectively. The following are examples:

    - `16` – Business Central 2020 wave 1

    - `17` – Business Central 2020 wave 2

    - `18` – Business Central 2021 wave 1

    - `19` – Business Central 2021 wave 2

    - `20` – Business Central 2022 wave 1

- **Second digit**: This gives us the minor release number. Minor releases come each month (almost). Typically, for SaaS environments, you have five minor releases after each major one.

- **Third digit**: This gives us the build number. The build number has five digits and it increases from update to update.

- **Fourth digit**: This gives us the revision number. This could have a `0` value (which means that it is the original release) or five digits such as a build (which means a hotfix was applied).

The version of my environment, `19.1.31886.32222`, could be decrypted as **Business Central 2021 wave 2, cumulative update 1, build 31886, hotfix 32222**.

- **Available update version**: This is not blank when an update for your environment is available and the number here shows the update version. You can schedule it on the environment's card; we will see this in the next chapter.

- **Scheduled update date**: When an update is available, you can choose any date and time in the **Schedule Environment Update** window of the environment card. You will see the chosen date in this field.

- **Security group**: This shows you the Azure Active Directory groups assigned to the environment. Only members of these groups can sign into the environment, even when they have a Business Central license:

  - **Not Set**: This is a default value, which means that any user with a Business Central license could sign in to the environment.

  - **Not Available**: This means that the previously assigned group is no longer available in the Azure Active Directory.

Creating a production or sandbox environment

In this section, you will learn how to create a new environment of both types, **Production** and **Sandbox**.

Production

By default, you can only have one production environment per tenant free of charge. By yourself, you could create a production environment only if you delete the existing one (please be very careful if you decide to delete your production environment). You can get more production environments in the following ways:

- Asking your CSP partner to create one more for you

- Asking Microsoft to migrate your environment from other tenants, if for some reason, you decided to merge them

Each additional production environment gives you three more sandboxes and an extra 4 GB of capacity:

1. To create a new production environment, click on **+ New** on the **Environments** list page:

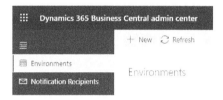

Figure 2.2 – New environment creation

2. In the **Create environment** window, type your new environment name.

3. Choose the **Production** type.

4. Select the needed country localization. If your country is not on the list, check the country availability here: `https://docs.microsoft.com/en-us/dynamics365/business-central/dev-itpro/compliance/apptest-countries-and-translations`. You might need to ask your partner to deploy it.

5. Choose the environment's version.

6.   After that, press the **Create** button.

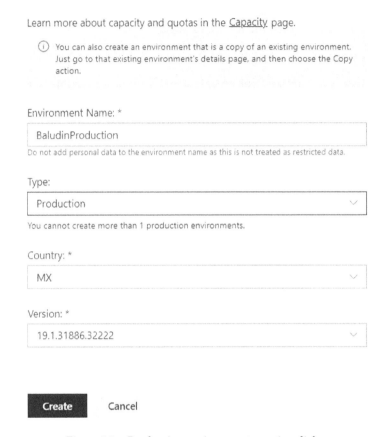

Create environment

Learn more about capacity and quotas in the Capacity page.

ⓘ You can also create an environment that is a copy of an existing environment. Just go to that existing environment's details page, and then choose the Copy action.

Environment Name: *

| BaludinProduction |

Do not add personal data to the environment name as this is not treated as restricted data.

Type:

| Production | ⌄ |

You cannot create more than 1 production environments.

Country: *

| MX | ⌄ |

Version: *

| 19.1.31886.32222 | ⌄ |

| **Create** | Cancel

Figure 2.3 – Production environment creation dialog

Important note

You can create a **W1** (World one) localization environment through a support request to Microsoft.

Sandbox

By default, you can only have three sandbox environments per tenant free of charge and you will get three more for each extra production environment.

The creation process is pretty much the same as for the production environment. Here, you just need to select the other environment type as **Sandbox**.

One moment! In the version field, you can find preview versions of future releases. Microsoft gives you the possibility to test the new functionality or your apps before the new version of Business Central is released. Just create the sandbox environment with a future release version. After it becomes generally available, your sandbox will be updated automatically.

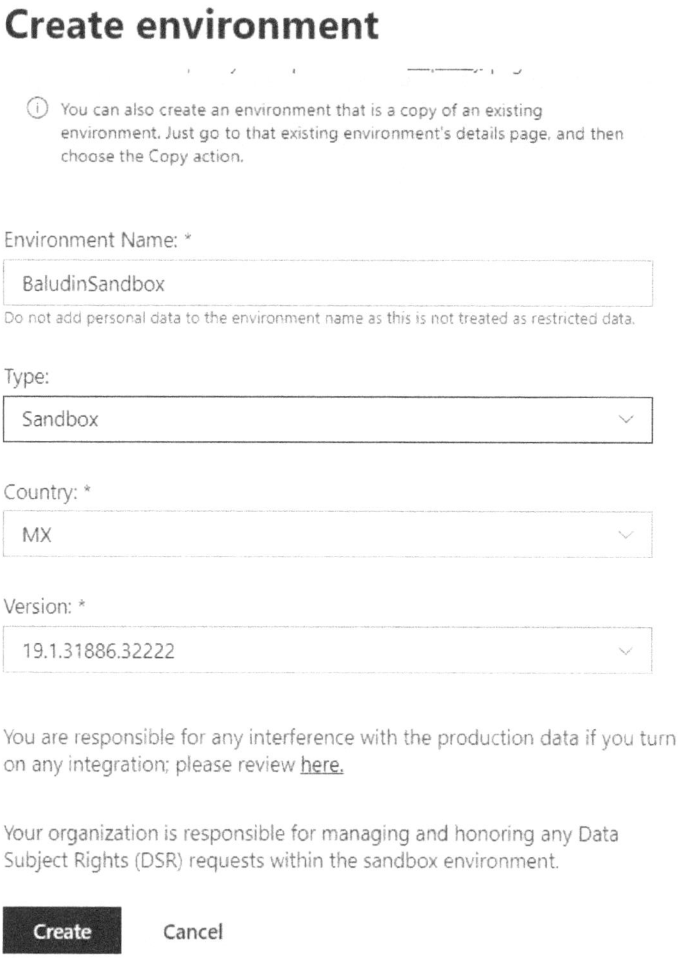

Create environment

ⓘ You can also create an environment that is a copy of an existing environment. Just go to that existing environment's details page, and then choose the Copy action.

Environment Name: *

> BaludinSandbox

Do not add personal data to the environment name as this is not treated as restricted data.

Type:

> Sandbox ⌄

Country: *

> MX ⌄

Version: *

> 19.1.31886.32222 ⌄

You are responsible for any interference with the production data if you turn on any integration; please review here.

Your organization is responsible for managing and honoring any Data Subject Rights (DSR) requests within the sandbox environment.

Create   Cancel

Figure 2.4 – Sandbox environment creation dialog

After you press **Create**, you will see your new environment on the list with the **Preparing** status:

Environments

Name	Application Family	Type	State	Country/region	Current Version
Production	Business Central	Production	Active	US	19.1.31886.32222
BaludinSandbox	Business Central	Sandbox	Preparing	DE	19.1.31886.32222
Sandbox	Business Central	Sandbox	Active	US	19.1.31886.32222

Figure 2.5 – Environment preparation

Now, wait until it changes to **Active**. Before this, you cannot use the environment. You could update the environment status by pressing the **Refresh** button:

Environments

Name	Application Family	Type	State	Country/region	Current Version
Production	Business Central	Production	Active	US	19.1.31886.32222
BaludinSandbox	Business Central	Sandbox	Active	DE	19.1.31886.32222
Sandbox	Business Central	Sandbox	Active	US	19.1.31886.32222

Figure 2.6 – Active environment

In the next section, we will see how to create a copy of your existing environment.

Creating a copy of the existing environment

Sometimes, you need to have a copy of the production environment. The reasons could be very different. Let's look at some of them:

- You might need a sandbox with fresh data.
- You might want to debug your production environment.
- You might want to test some new features.

For all of these cases, you would like to have a production copy, or maybe some other environment copy.

Using this functionality, you can copy not only production environments but also any other environment that you have. You can do this quickly with a few clicks. Even if after your tests, data is crushing in your environment copy, just delete it and create a copy again.

So, let's see what you need to do if you want to copy your environment:

1. Open the **Environments** list and click on the environment's name that you want to copy.

2. On the environment's card, click on the **Copy** button:

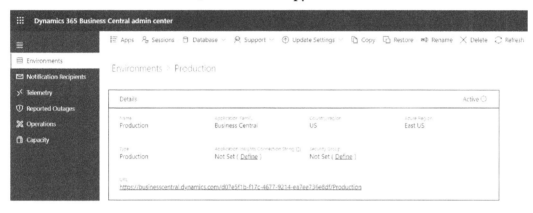

Figure 2.7 – Environment card

3. In the **Copy environment** window, type your new environment's name and select the new environment type:

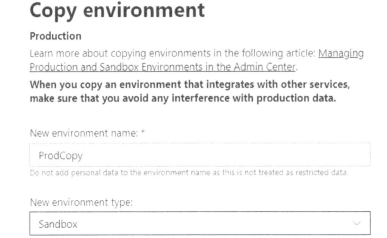

Figure 2.8 – Copy environment dialog

Remember the environment limitations. If you try to create a new production environment and you already have one, and you did not buy another one through your CSP partner, then you will get the following error message:

Figure 2.9 – Environment creation error

4. Now, click on the **Copy** button and confirm the environment's creation:

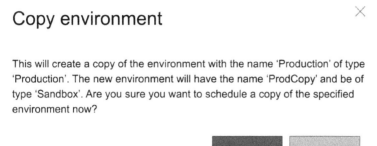

Copy environment   ✕

This will create a copy of the environment with the name 'Production' of type 'Production'. The new environment will have the name 'ProdCopy' and be of type 'Sandbox'. Are you sure you want to schedule a copy of the specified environment now?

Yes   No

Figure 2.10 – Copy environment confirmation

Wait until your environment status changes from **Preparing** to **Active**, and you can then enjoy your fresh copy.

> **Important Note**
>
> If you have some integration connections in your production environment, then disable them in the production copy; otherwise, you will get test data in your integration.

5. You can rename and delete your environments using the **Rename** and **Delete** options. These actions are placed on the environment's card (not list). Just select the needed action and confirm it:

Figure 2.11 – Rename and Delete environment actions

> **Important Note**
>
> Environment renaming will change the environment's URL. All integrations that use any links to the environment will be broken and you will have to update them. Also, renaming causes the environment to restart and all users will be signed out.

With this, you will get the following dialog boxes for your respective selection:

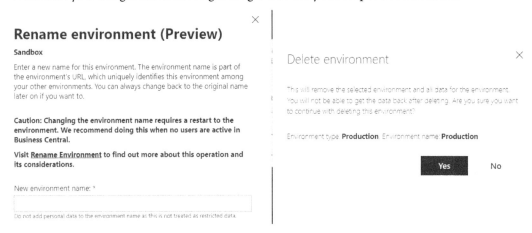

Figure 2.12 – Rename and Delete dialogs

Summary

You now know everything about the **Environments** list, so you can easily get an overview of your environments. You know how to create a new environment of both types. You can create a copy of any environment for your purposes. And finally, you can rename and delete your existing environment.

Now, you are ready to deep dive into environment management.

3
Environment Details and Notifications Setup

Environment management is not only limited to creating, copying, and deleting functions. You can set up access to your environments, create and restore backups, schedule updates, and many more things. In addition, you can set up notifications of some events to your email to avoid opening the Admin Center every day, waiting for an environment update, for example.

In this chapter, we are going to cover the following main topics:

- Detailed information about your environment
- Security group setup
- Notifications setup
- Upgrade scheduling
- Apps and sessions

- Database export
- Backup restore

After this chapter, you will be able to perform full environment management, manage updates, restore backups, and create database copies.

Detailed information about your environment

Detailed environment information is hidden in the **environment card**. You can open it by clicking on the environment's name in the **environments list**.

The environment card looks like this:

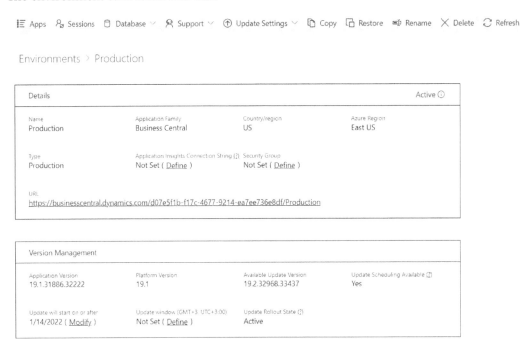

Figure 3.1 – The environment card

The card has the following sections.

Action buttons

These are placed on the top of the card and used for different things you can do with the environment. In this section, you will get a short introduction, and all the actions will be explained in detail in the next sections. You have the following actions available:

1. **Apps**: This is a package of functionality. The standard one is provided by Microsoft; and the additional ones are provided by you or third parties. This consists of the list of apps installed in your environment, with names, publishers, versions, and available updates. This list does not belong to the only chosen environment – you can choose any of your environments at the top-left corner:

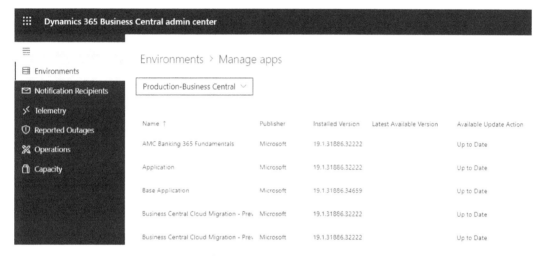

Figure 3.2 – Manage apps

2. **Sessions**: This contains a list of user sessions with some additional details. Like the **Apps** action, you can select the needed environment directly from the list. You can also restart your environment here:

Figure 3.3 – Manage sessions

3. **Database**: This consists of two sub-actions:

    - **Create database export**: To export your database to an Azure storage container.

    - **View export history**: To view the export logs. Database export is only available for **production** environments and only if you have paid for a Business Central license:

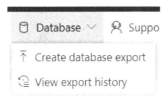

Figure 3.4 – Database action menus

4. **Support**: This consists of three sub-actions that allow you to manage your support requests:

    - **New Support Request** creates a common issues ticket.

    - **Report Production Outage** creates an outage ticket for serious incidents such as a block of all users signing in or an API's inaccessibility.

    - **Manage support contact** allows you to fill the primary support contact for the selected environment.

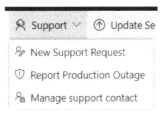

Figure 3.5 – Support action menus

5. **Update settings**: This consists of two sub-actions that allow you to manage your environment updates:

    - **Set update window**: You can set time intervals when your environment can be updated (a minimum of 6 hours required).

    - **Set update date**: You can select the update day.

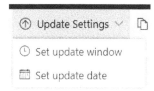

Figure 3.6 – The Update Settings action menus

6. **Copy**: This is used for environment copy creation.

7. **Restore**: This is used for database backup restoration.

8. **Rename**: This is used for environment renaming.

9. **Delete**: This action is used to delete your environment.

10. **Refresh**: This action is to update information on the environment's card – for example, to track a status change.

> **Note**
> The **Copy**, **Rename**, and **Delete** functions are covered in detail in *Chapter 2*.

Details

This tab contains common environment information:

1. The current environment's status.

2. The name of the selected environment.

3. **Application Family** – it is Business Central for now.

4. **Country\region** – the localization version of your environment.

5. **Azure region** – where your environment is placed in the Azure cloud.

6. **Type** – the environment's type (Production or Sandbox).

7. **Application insights connection string** – used for telemetry setup. We will dive into this deeply in *Chapter 4*.

8. **Security Group** – you can assign an existing security group from Azure to the selected environment, and only users from this group will be able to sign in. Administrators are able to sign in any way. All users must have a Business Central license (except the delegated admin). We will look at the security group setup later in this chapter.

9. **URL** – your environment URL. You can copy this value from here if you cannot construct the environment's URL by yourself.

Version management

This contains information about your current version and the available update:

1. **Application version** – the current version of your **base application**.

2. **Platform version** – the current version of your platform.

3. **Available update version** – the version of your base application after the update.

4. **Update scheduling available** – shows whether you can reschedule an update or not. The possible values are **Yes** and **No**.

5. **Update will start on or after** – the possible update date. When updates become available, you will see a default update date here. You could change it if the value of **Update Scheduling available** is **Yes**.

6. **Update window** – the time interval when an update will apply. You can also change it if it is possible. The interval must be 6 hours as a minimum.

7. **Update rollout state** – sometimes, Microsoft can discover some critical issue in the available update. In this case, you will see that this field has a **Postponed** value. Personally, I have only encountered such a situation for major updates. In addition, you will be notified to your email and in the Admin Center. Just relax and wait until Microsoft resolves the issue, and you will be available to reschedule an update or get it automatically.

Now that you know everything about each element from the environment card, we can learn more about processes that you can perform. Let's start with the **security group** setup in the next section.

The security group setup

In this section, you will learn how to restrict access to your environments with a simple security group setup. But why would you want to restrict access to the environment? Usually, this happens when you have created an environment for some specific operations. This could be a production or sandbox environment. Maybe you don't want all users to have access to this environment's data or you want to test something new and prevent users from creating operations here. Whichever way, we can set up a limitation policy.

Let's look at the setup steps:

1. First of all, open the **Azure portal** (`https://portal.azure.com/`) and search for `groups`:

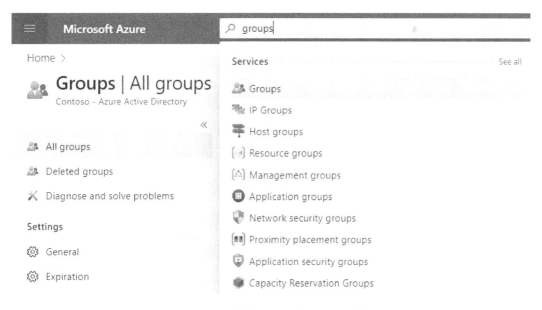

Figure 3.7 – The Azure Groups search

2. After that, click on the **New group** action to create a new security group:

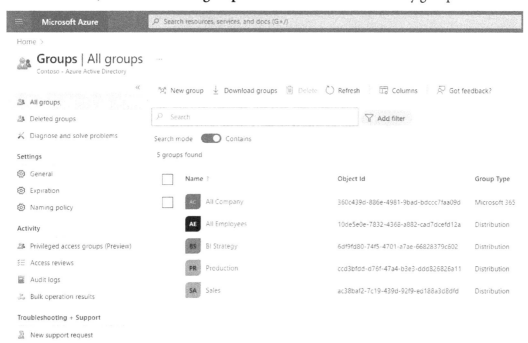

Figure 3.8 – The Azure Groups list

3. Set **Group type** as **Security**, and input the group name and group description. Let's name it `Business Central Production`. After that, click on **Owners** and **Members**, add existing users or existing groups as group members, and finally, click on the **Create** button:

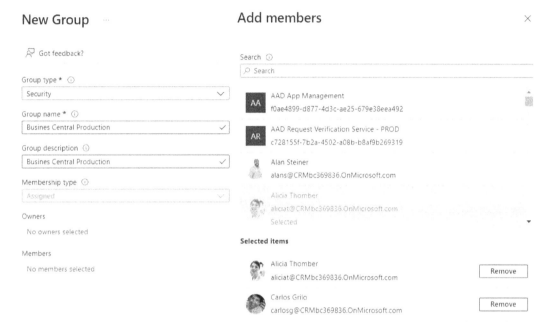

Figure 3.9 – The New Group dialog

4. Now, turn back to the environment card in the Admin Center and click on the **Security Group | Define** link. You will see the next dialog, where you can select a created security group. Click on the **Save** button, and the security group will be assigned to your environment:

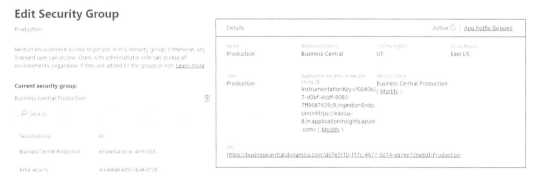

Figure 3.10 – The Edit Security Group dialog

Up until this moment, only users who have a Dynamics 365 Business Central license and belong to the **Business Central Production** security group can sign in to the environment (and delegated admins, of course). To remove the security group, just click on the **Modify** button near the group name, press the **Delete** icon near your group, and click on **Save**:

Edit Security Group

Production

Restrict environment access to people in this security group. Otherwise, any licensed user can access. Users with administrator role can access all environments, regardless if they are added to the group or not. Learn more

Current security group:

Business Central Production   🗑

🔍 Search

SecurityGroup	Id
Business Central Production	e638ad3a-ec5e-4b90-b58...

Figure 3.11 – Deleting the security group

Now that you know how to restrict access to the environment using Azure security groups, we can move on to how to find out that an update for your environment is available.

Notification setup

In this and the next sections, we will look at cloud environment updates and how to prepare and schedule them. But first of all, we need to know that an update is available. Opening the Admin Center and checking for updates every day is not a practical way. So, let's take a look at functions such as **notification recipients**.

> **Important Note**
> It is highly recommended to set up notification recipients just after you create your first environment to prevent an unexpected update.

Notification recipients

1. Open the Admin Center and click on the **Notification Recipients** tab. Then, click on the **+ Add recipient** button:

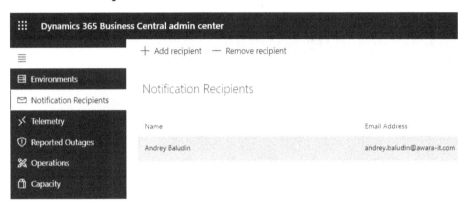

Figure 3.12 – Notification Recipients

2. Then, fill in the recipient's name and email address, and press **Save**:

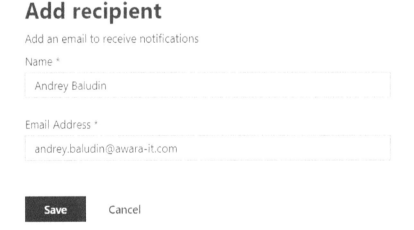

Figure 3.13 – The Add recipient dialog

These are the following notifications that the selected account will get after a notification setup:

- A major or minor update is available for an environment.

- An update is scheduled.

- The environment was successfully updated.

- The environment was not updated.

- An environment update failed.

- There are update conflicts with some of the installations in the environment apps.

All recipients from the notification list will get the same notifications, and they will get notifications about all the environments from the list.

Here are notification samples from my practice. The update appeared, and the environment was updated successfully:

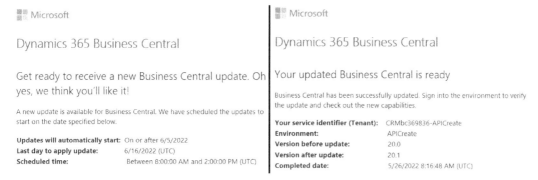

Figure 3.14 – A successful update notification

An update can fail for different reasons – an issue on the Microsoft side, an issue in your environment, and issues with your extensions:

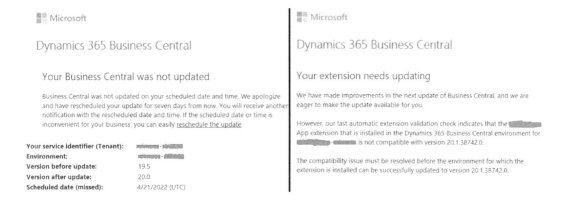

Figure 3.15 – Failed update notifications

> **Important Note**
> Microsoft uses the Microsoft Partner Center `msftpc@microsoft.com` account to send notifications. Don't forget to check that emails from this sender are not being sent to your spam folder.

Don't forget to remove email from notification recipients if you stopped supporting the customer or chose another partner.

Now that you know how to be notified about future environment updates, we can move on to learn how to schedule an update.

Upgrade scheduling

As we know from previous chapters, there are two update types that you can schedule:

- Major updates – twice a year in April and October
- Minor updates – usually on the first Friday of the month, except for April and October

So, you've got a notification that a new update is available. What are your next actions?

> **Important Note**
> If you do not set the upgrade date, your environment will be updated on any day between the default date and the last upgrade date. This can cause an interruption of a normal user's work.

Environment updates are managed by Microsoft, and all that you need to do is schedule them. Before you schedule your production environment update, I would advise you to do the following things:

- Set an update date of the production environment in the future for two reasons:

    I. I want to ensure that my per tenant extensions are fully compatible with the incoming update, so I will update the sandbox environment first to check this.

    II. I do not want to get a production environment update in the default date and time window because it could be during business hours. So firstly, I will set the update date as far in the future as possible, and secondly, after a successful sandbox update, I will reschedule the production update to a suitable date and time.

- I will update one of my sandbox environments first to check that my existing per tenant apps are compatible with the update (if you have not done this before, use a future release **Docker** image – this is not a topic for this book, but is worth mentioning). For this, you need to perform the following actions:

I.   Open your Sandbox environment's card and click on the **Modify** link near the **Update will start on or after** field, or click on the **Update Settings** button and select **Set Update Date**. Select the nearest available update date.

II. Click on the **Define** link near the **Update window** field or choose **Update Settings | Set Update Window**. Select a 6-hour update window. Usually, I select a time interval at night to prevent user work issues.

Important Note

Pay attention to your time zone if you set up an update window for your customer. It is shown near the **Start Time** and **End Time** fields. In case of a wrong update window setup, the environment may be inaccessible during customers' work hours.

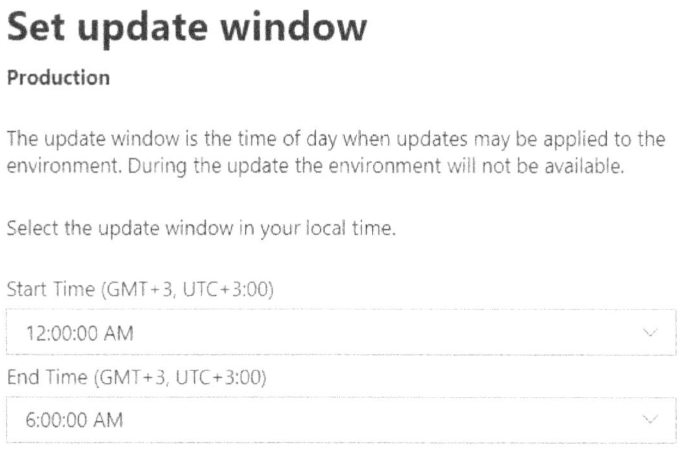

Set update window

Production

The update window is the time of day when updates may be applied to the environment. During the update the environment will not be available.

Select the update window in your local time.

Start Time (GMT+3, UTC+3:00)

12:00:00 AM	⌄

End Time (GMT+3, UTC+3:00)

6:00:00 AM	⌄

Figure 3.16 – Set update window

III. After that, you will get an email notification (if you did the notification setup before) that your update is scheduled. Now, wait until you will get an email notification that your updated sandbox environment is ready.

Important Note

All of your apps that were published to the sandbox from Visual Studio Code, will be uninstalled due to the environment's update.

IV. After a successful sandbox environment update, you need to check that everything is okay with your apps. Download new symbols for your app and fix new errors and warnings if they appear.

V. When everything is checked, you can schedule the production environment update date and time window the same way you did with the sandbox environment.

Possible upgrade issues

So, you got a notification that your upgrade failed. Let's look at the possible reasons for this:

•   Your per tenant extension is not compatible with the major update.

    Solution: Fix your app and deploy it to the next major version:

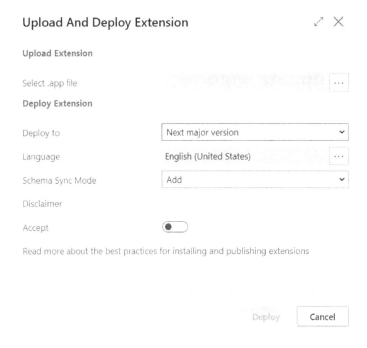

Figure 3.17 – Upload extension

•   Your installed AppSource app is not available for the next major update.

    Solution: If this app is not important, you can just uninstall it and reschedule the upgrade. If it is important, you need to contact the app developers for details.

- You have an active cloud migration.

  Solution: Ask your partner about cloud migration activity, then open the **Cloud Migration Management** page in Business Central, and click on **Disable Cloud Migration**:

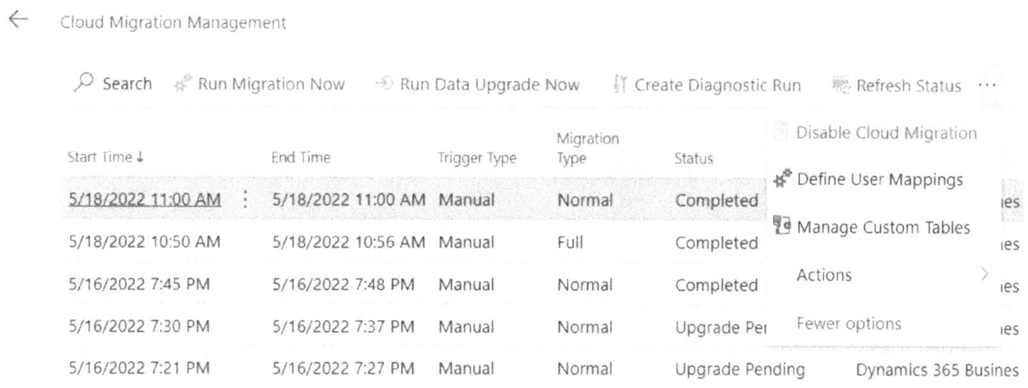

Figure 3.18 – Disable Cloud Migration

- Your environment doesn't have the latest update installed – for example, you're trying to install a major update of version 19.0 and your environment has version 18.4. Minor update 18.5 exists and you postponed its installation for a long time.

- **Solution**: Reschedule the upgrade. Do not postpone upgrades until the last moment.

- Microsoft can postpone the upgrade if some critical issues are discovered. It happens mostly with major updates. You will get a notification of the postponement issue, and your environment will get an upgrade schedule later without a definite date.

  You can prepare for a major update by installing early some new features from the **Feature management** page of Business Central. Select the available feature and choose **All users** in the **Enabled for** field. The data upgrade will run automatically.

Remember that you cannot turn off the enabled feature, so test this in a sandbox first:

Figure 3.19 – Feature Management

Now, you are able to manage and schedule updates and know how to fix upgrade issues. We will now move on to **apps** and **sessions** management.

Apps and sessions

The Admin Center helps you to manage many things related to your environments. This includes Microsoft standard apps and apps from **AppSource** that you have installed in your environment:

1. Open an environment card and click on the **Apps** button at the top. You will see a list of your apps (per tenant extensions are not included):

Environments > Manage apps

Production-Business Central ∨

Name ↑	Publisher	Installed Version	Latest Available Version	Available Update Action
AMC Banking 365 Fundamentals	Microsoft	20.1.39764.40432		Up to Date
Application	Microsoft	20.1.39764.40432		Up to Date
Base Application	Microsoft	20.1.39764.41988	20.1.39764.42040	Install update
Business Central Cloud Migration - Prev	Microsoft	20.1.39764.40432		Up to Date

Figure 3.20 – The apps list

Here, you can see that one app from AppSource has an available update. If you click on **Install Update** button, the status will change to **Update is scheduled**, and within some time, the update will install to your environment. If you did not apply app updates, work with the old version of the app until the next major environment update, when it will apply automatically:

Base Application   Microsoft   20.1.39764.41988   20.1.39764.42040   Update is Scheduled

Figure 3.21 – Apps update

2. Sometimes, you can see the **Button required** status – click on it and read what you need to do to upgrade this app. Usually, you need to update or install some dependent app first.

3. You don't need to reopen an environment card to see the other environment-installed apps – just select the needed environment directly from the **Manage apps** list:

Figure 3.22 – The Manage apps environments

4. Now, let's learn how to manage user sessions. For this, open the environment card again and click on the **Sessions** button. You will see the **Manage sessions** list:

Figure 3.23 – Manage sessions

Here are all the users who now work with your environment. The list contains just the most important information in the shortest form: **Session ID**, **User ID**, **Client Type**, and **Login Duration**. You can get detailed session information by clicking the **Show session details** checkbox.

You can refresh the sessions list by pressing the **Refresh** button at the upper-right corner.

If something went wrong with the user's session (or sessions) or you need just to log off everybody from the environment, you can press the **Cancel session** button near each session (even your own), or select multiple sessions and press the **Cancel selected sessions** button at the top of the list.

The last important option here is the **Restart Environment** button. You may need to perform this in some urgent cases. Just press the button and confirm it:

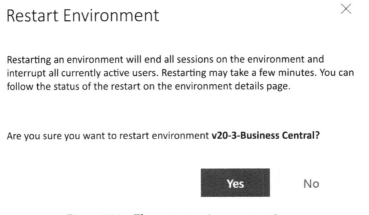

Figure 3.24 – The restart environment option

Important Note
All users will be kicked off on a restart. Send them a restart warning before. In addition, check your jobs and integration status after the restart.

In addition, as in the **Manage Apps** functionality, you are able to select environments directly from the **Manage sessions** list and don't need to open the environment card for this.

Database export

You may want to create your cloud database export – for example, to deploy extra test environments, provide database copy to some third party, or even migrate from SaaS to on-premises.

How do you export your database if it is somewhere in the cloud? This is a very important question, which we will discuss in this section.

First of all, let me notify you about some database export limitations:

- You are able to export databases only for a **production** environment.

- You are able to perform 10 exports per month.

- You must be delegated admin or admin in the current organization. Your Business Central user must have the **D365 BACKUP\RESTORE** permission set assigned.

- Your Dynamics 365 Business Central license must be paid.

Now, we can start our database export. You are able to export databases to Azure Storage accounts only. This means that you must have a paid or trial Azure subscription to perform this button. If you don't have a suitable account, you can create a new one:

1. Open the Azure portal and find **Storage Accounts**.

2. Click on the **+ Create** button. Choose your subscription, create a new resource group, and input the storage account name and region. Performance must be **Standard**. Press **Review + Create** and, after that, **+ Create** one more time. Wait until your new storage account is ready:

Project details

Select the subscription in which to create the new storage account. Choose a new or existing resource group to organize and manage your storage account together with other resources.

Subscription *	Visual Studio Enterprise Subscription – MPN	⌄

Resource group *	(New) BCExport	⌄
	Create new	

Instance details

If you need to create a legacy storage account type, please click here.

Storage account name ⓘ *	bcexporttest2

Region ⓘ *	(US) East US	⌄

Performance ⓘ *	⦿ **Standard:** Recommended for most scenarios (general-purpose v2 account)

Figure 3.25 – Creating the storage account

3. Now, you need to generate a **Shared Access Signature (SAS)** to provide access to your container. Open your storage account and choose **Shared access signature**. Under **Allowed services** choose **Blob**, under **Allowed resource types** choose **Container** and **Object**, and under **Allowed permissions** choose **Read**, **Write**, **Delete**, and **Create**. Set the **Start and expiry date\time** (at least 24 hours). I recommend that you set the period in which you're planning to work with database exports. Click **Generate SAS and connection string**:

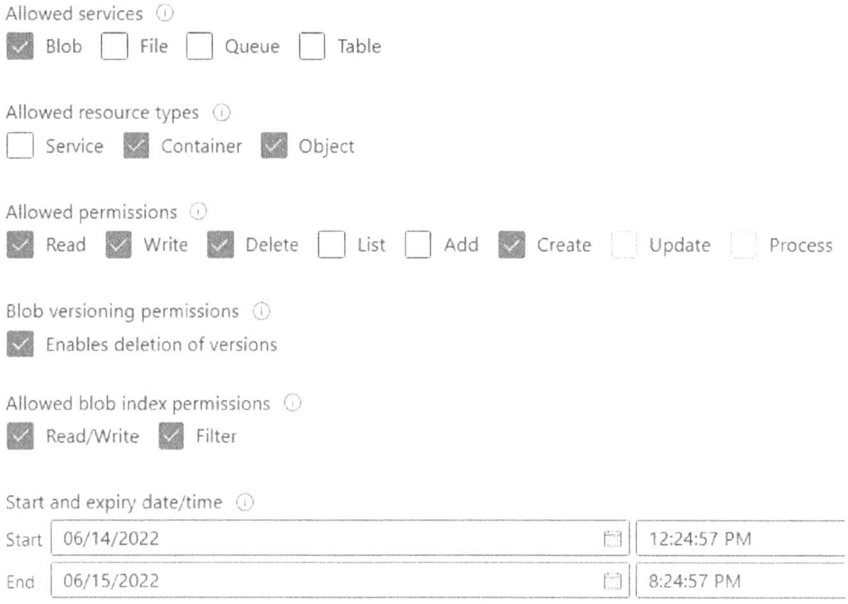

Figure 3.26 – Creating a shared access signature

4. You will see the next values automatically generated. Copy the link in the **Blob service SAS URL** field.

Signing key ⓘ

key1

Generate SAS and connection string

Connection string

BlobEndpoint=https://bcexporttest.blob.core.windows.net/;QueueEndpoint=https://bcexporttest.queue.core.windows.net/;FileEndpoint= ... ⓓ

SAS token ⓘ

?sv=2021-06-08&ss=b&srt=co&sp=rwdctfx&se=2022-06-15T17:24:57Z&st=2022-06-14T09:24:57Z&spr=https&sig=gbKNB5xt490favtli ... ⓓ

Blob service SAS URL

https://bcexporttest.blob.core.windows.net/?sv=2021-06-08&ss=b&srt=co&sp=rwdctfx&se=2022-06-15T17:24:57Z&st=2022-06-14T09:... ⓓ

Figure 3.27 – Generate SAS

5. Open the Admin Center and go to the environment card. Click **Database** and **Create Database Export**. Fill the **SAS URI** field with the SAS URL that you just copied. Change the container name if you don't like the default name:

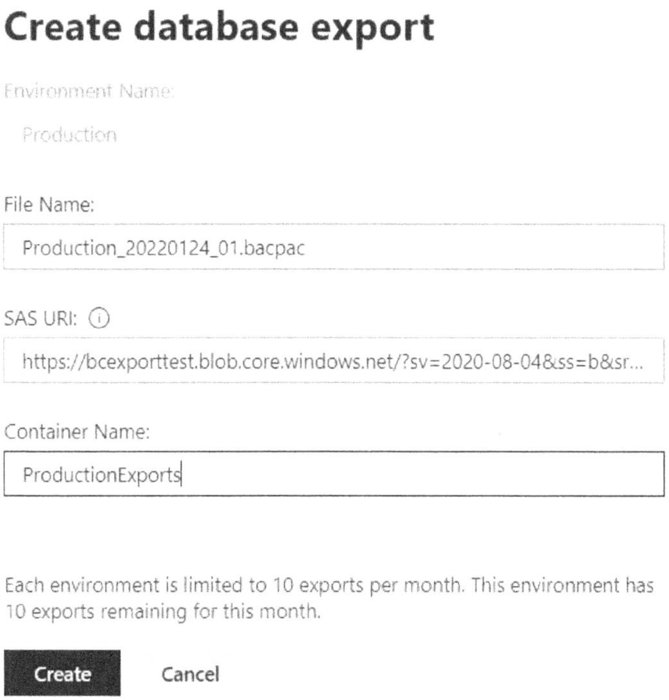

Create database export

Environment Name:

Production

File Name:

Production_20220124_01.bacpac

SAS URI: ⓘ

https://bcexporttest.blob.core.windows.net/?sv=2020-08-04&ss=b&sr...

Container Name:

ProductionExports

Each environment is limited to 10 exports per month. This environment has 10 exports remaining for this month.

Create   Cancel

Figure 3.28 – Creating a database export

6.   Press **Create**, and you will get a confirmation message:

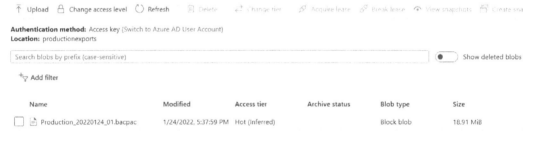

Figure 3.29 – The database export confirmation

Depending on your database size after some time, you will see your database as
a blob inside the storage account's container:

Figure 3.30 – Exported database

7.   You can check the export history if you open the environment card from the Admin
Center and click **Database** and **View export history**:

Environments > Export History

Date/Time	Environment	Application Family	Country	Version	Action	Export name	Storage
1/24/2022 5:26:28 PM	Production	Business Central		19.2.32968...	Create database export	Production_20220124_01.bacpac	bcexpor

Figure 3.31 – Export history

You can use a created export to restore the database on your SQL server or Docker container environment. We will learn how to restore your cloud environment in the next section.

Backup restore

Sometimes, things go totally wrong and your database is crushed, or you need to roll out some situation. Or maybe everything is okay and you just need to refer to some earlier data. In both cases, we can use database restore. Also, the good news is that you don't need to take care of backup policies manually using the Business Central SaaS. Backups are created automatically, and you just need to choose a point of restoration.

But not everything is ideal; there are some limitations:

1. You can restore your backups up to 28 days in the past.
2. You can restore your environment only to the same version you have now. So, after each minor or major update, your 28-day restoration period shrinks to 0 and starts again.
3. You need the same permissions as a database export.
4. Restoration is limited to 10 times per month.
5. You can't restore a deleted environment from the Admin Center – contact Microsoft support for this.
6. Sandbox environments can be restored only as sandbox environments. Production environment restoration supports both types.

The restoration process looks very simple:

1. Make sure that you don't exceed the environment quota. If you do, delete the unneeded environment.
2. Open the environment card and click the **Restore** button.

3. Choose the backup date, time, and environment type, and input the environment's name. Click **Restore** to continue, wait until your environment is ready, and then change the status from **Preparing** to **Active**:

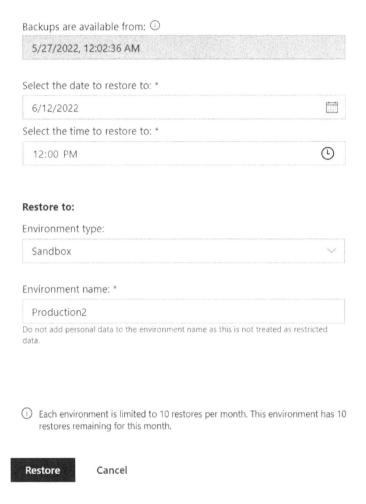

Figure 3.32 – Restore environment

4. After that, you can delete the old environment if you don't need it, and rename the restored environment with the old environment's name.

5. Here are some extra checks for the restored environment:

    -   Stop all the jobs in the restored environment until you've checked everything.

    -   If you restored the sandbox environment with extensions published from Visual Studio Code, you will need to publish them again, as after the environment update. Uploaded apps will stay installed.

    -   Some things such as CRM connections and Dataverse connections will be disabled, and you will need to create them again. Power BI's incremental refresh will require a reset.

Summary

In this chapter, we covered a lot of situations that you can encounter as a Business Central administrator. You learned how to get environment details, limit access to the environment, and set up notifications about available updates and schedule them. You now know how to update installed apps, restart a cloud environment, and manage user sessions. Finally, you learned important things such as how to export your database and restore a backup.

In the next chapter, we will move on to telemetry analysis.

4
Telemetry Setup and Analysis

What is telemetry? It is an automated communication process from multiple data sources. Telemetry helps you to understand what's going on with your software – Does anybody use it? How do they use it? And if something went wrong, then when and why did that happen? In some cases, you will know about the problem even earlier than your customer.

	timestamp [UTC]	message	severityLevel	itemType	customDimensions	operation_Name
>	2/9/2022, 10:56:27.179 A...	Operation exceeded time threshold (SQL q...	2	trace	{"aadTenantId":"[redacted]","co...	Long Running O
>	2/9/2022, 10:22:05.519 A...	Operation exceeded time threshold (AL m...	2	trace	{"extensionVersion":"19.3.34541.35754","extensionId":"437dbf0...	Long running op
>	2/8/2022, 11:31:06.156 AM	Operation exceeded time threshold (SQL q...	2	trace	{"component":"Dynamics 365 Business Central Server","enviro...	Long Running O
>	2/8/2022, 3:21:03.558 PM	Operation exceeded time threshold (AL m...	2	trace	{"companyName":"[redacted]","component":"Dynamics 365 ...	Long running op
>	2/8/2022, 3:21:03.558 PM	Operation exceeded time threshold (AL m...	2	trace	{"companyName":"[redacted]","component":"Dynamics 365 ...	Long running op

Completed. Showing results from the last 24 hours.   00:01.1   5 records

Figure 4.1 – Telemetry sample

When you report to Microsoft regarding some issues with your environment, they always check the telemetry because it will tell them more than you will.

To be clear, you will be familiar with **Change Log** in Business Central. It is a table where system logs record changes, provided you have set up it before. Telemetry behaves like that. It logs a user's actions and environment events and sends them to cloud storage. You have a set of default events that are being logged automatically, and you could collect the telemetry using code. It is a record in the log of an event. Detailed or not, it depends.

Environment telemetry in the Admin portal has quite limited opportunities in terms of analysis, so in this chapter, we will use some extra tools for this process. The following topics will be covered:

- Environment telemetry in the Admin portal

- Environment telemetry in Application Insights

- Analyzing telemetry

After this chapter, you will be able to set up the telemetry on a tenant or extension level and analyze it with the different tools.

> **Important Note**
>
> There is an official Microsoft repository on GitHub with multiple telemetry samples at `https://aka.ms/bctelemetrysamples`.

Environment telemetry in the Admin portal

We will have an initial encounter with the telemetry and you will understand how to analyze it using the Admin portal.

Here, in the **Admin portal**, you can find only top-level events from your environments with filters by date and time. Detailed telemetry is available in **Application Insights** and we will get to know this in the sections that follow.

First of all, open the Admin portal and click on the **Telemetry** tab.

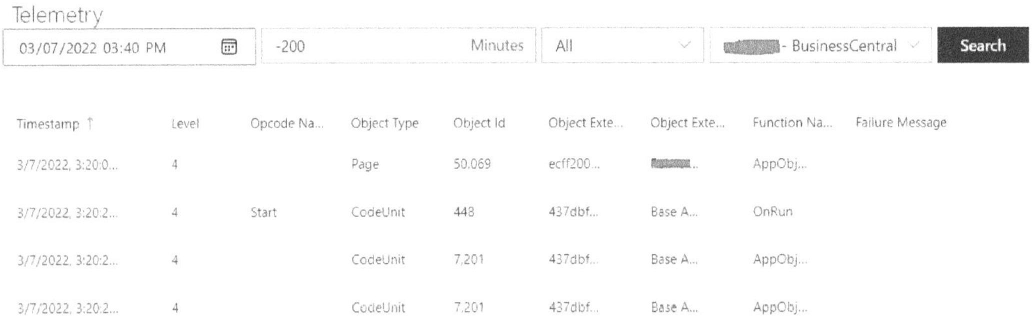

Figure 4.2 – Telemetry at the Admin portal

To obtain the telemetry, perform the following steps:

1. In the last field, choose the environment required.

2. In the first field, set the start date and time point from which you want to get the telemetry – ask the user about the issue's approximate time and set it here.

3. In the second field, set a time interval in minutes: just set the + or – quantity of minutes.

4. Press the **Search** button and you will see the telemetry records.

> **Important Note**
>
> The Admin portal telemetry contains only top-level AL events and any errors from calls through the telemetry stack.

The **Telemetry** tab consists of these columns:

1. **Timestamp** of the event

2. (Severity) **Level**: contains the number of levels:

Level	Description
1	Critical
2	Error
3	Warning
4	Normal
5	Verbose

3. **Opcode name**: **Start** or **Stop**

4. **Object type**: AL object type (table, page, code unit, and so on)

5. **Object Id**: AL object ID

6. **Object Extension Id**: ID of the app where the event was raised

7. **Object Extension Info**: Name of the app

8. **Function Name**: This could contain trigger names such as **OnRun** or **OnInit**, or could contain the call stack, for example:

    ```
    AppObjectType: CodeUnit AppObjectId: 1310
    AL CallStack:
    "O365 Activities Dictionary"(CodeUnit 1310).
    OnRun(Trigger) line 12 - Base Application by Microsoft
    ```

9. **Failure Message**: Contains the error text if an error has occurred

You can sort the telemetry records by clicking on the column name, for instance, to get records with a higher severity level.

Unfortunately, the filtering and selection possibilities offered by the Admin portal telemetry are quite poor, so in the next section, we will learn how to use advanced telemetry in **Application Insights**.

Environment telemetry in Application Insights

Application Insights is a feature of **Azure Monitor**, which provides you with the possibility to monitor application performance and usage. If you want to use it, you need to create this resource in the **Azure** (not Admin) portal and link it to your environment or extension.

> **Important Note**
> You have only 5 GB of free telemetry data per month. Every extra gigabyte of data will cost you approximately USD 2–3 per month. Remember that some events, such as HTTP calls, could create a lot of telemetry data (several gigabytes per day), but you can always limit it in Application Insights. If you are a partner, and combining telemetry from different customers in **Application Insights**, you will exceed your free limit very quickly.

To begin your work with Application Insights, you need to create it as a resource in the Azure portal. Also, you must have an active Azure subscription:

1. Open the **Azure portal** (`https://portal.azure.com/`) and find **Application Insights** there.

Figure 4.3 – Finding Application Insights

2. Choose your subscription and complete the fields. You can check the actual price here: `https://azure.microsoft.com/en-us/pricing/details/monitor/`. Microsoft recommends choosing the region near you or where your Business Central is hosted in order to maximize performance.

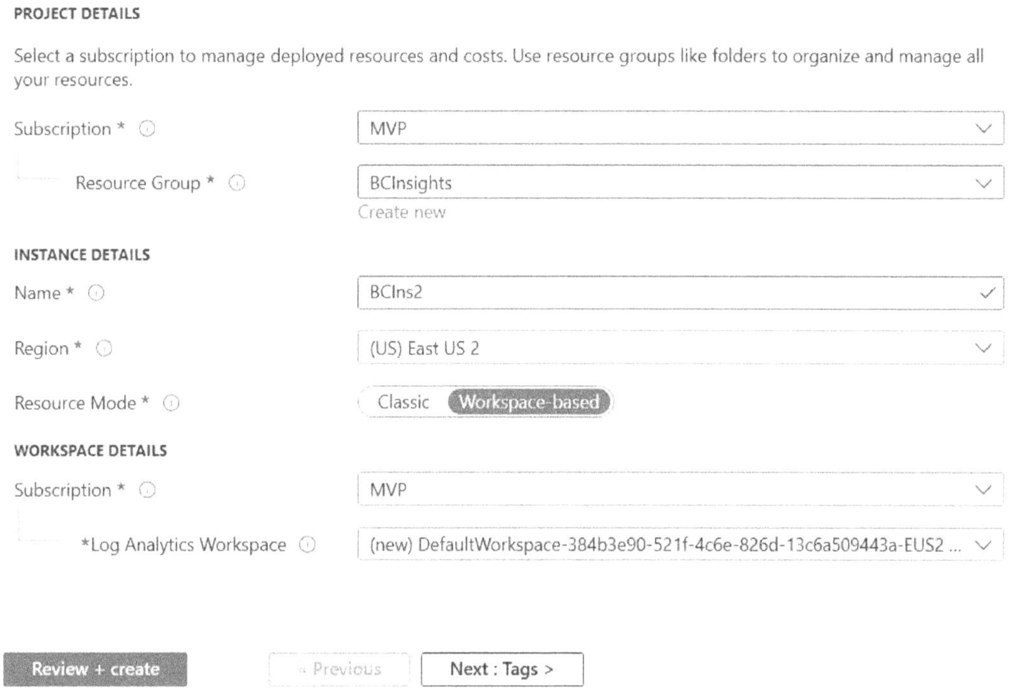

PROJECT DETAILS

Select a subscription to manage deployed resources and costs. Use resource groups like folders to organize and manage all your resources.

Subscription * ⓘ	MVP ⌄
Resource Group * ⓘ	BCInsights ⌄
	Create new

INSTANCE DETAILS

Name * ⓘ	BCIns2 ✓
Region * ⓘ	(US) East US 2 ⌄
Resource Mode * ⓘ	Classic **Workspace-based**

WORKSPACE DETAILS

Subscription * ⓘ	MVP ⌄
*Log Analytics Workspace ⓘ	(new) DefaultWorkspace-384b3e90-521f-4c6e-826d-13c6a509443a-EUS2 ... ⌄

[Review + create]   ‹ Previous   [Next : Tags >]

Figure 4.4 – Creating Application Insights

3.   Click on **Review + create** and wait until validation is complete.

Figure 4.5 – Application Insights validation

4.   Click on **Create** again.

5.   Open the Application Insights **Overview** field and copy a connection string from there.

Figure 4.6 – Application Insights overview

6.   Now you can apply this connection string to collect the environment's telemetry. Open your environment card in the Admin portal and click on the **Define** link near the **Connection String** field. Set the **Enable Application Insights** mark and paste your connection string. Finally, click **Save**.

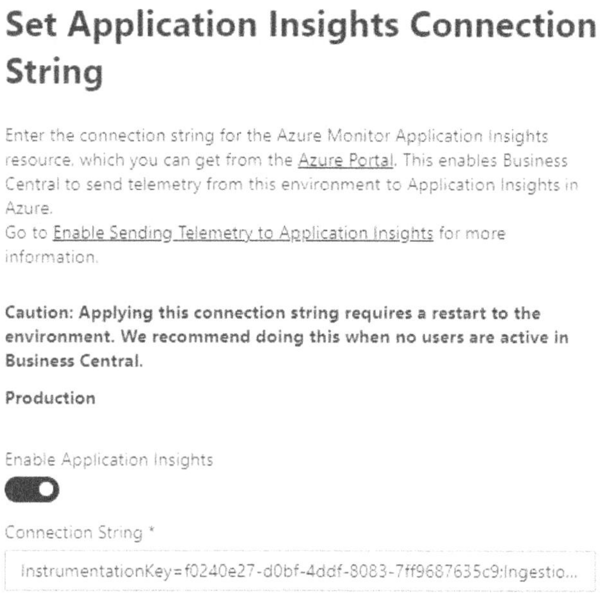

Set Application Insights Connection String
Enter the connection string for the Azure Monitor Application Insights resource, which you can get from the Azure Portal. This enables Business Central to send telemetry from this environment to Application Insights in Azure.
Go to Enable Sending Telemetry to Application Insights for more information.
Caution: Applying this connection string requires a restart to the environment. We recommend doing this when no users are active in Business Central.
Production
Enable Application Insights
Connection String *
InstrumentationKey=f0240e27-d0bf-4ddf-8083-7ff9687635c9;Ingestio...

Figure 4.7 – Set Application Insights Connection String

> **Important Note**
>
> Your environment will revert to the **Preparing** status and will be inactive and won't restart for some time. Users might be kicked off. Please do not set this up during business hours.

7. After all the actions, you will see your connection string filled and your environment must have a status of **Active**. You must repeat all these steps for each environment you want to monitor.

Environments > Production
Details
Active
Name
Production
Application Family
Business Central
Country/region
US
Azure Region
East US
Type
Production
Application Insights Connection String
InstrumentationKey=f0240e27-****-****-****-************;IngestionEndpoint=https://eastus-8.in.applicationinsights.azure.com/ (Modify)
Security Group
Business Central Production (Modify)

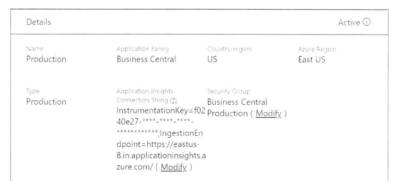

Figure 4.8 – Application Insights in the Environments card

8. The other possibility is to set up the telemetry for your extension and get the telemetry to **Application Insights** to be aware of issues earlier than the customer. To do this, add the `applicationInsightsConnectionString` element to the `app.json` file of your extension and paste the connection string there.

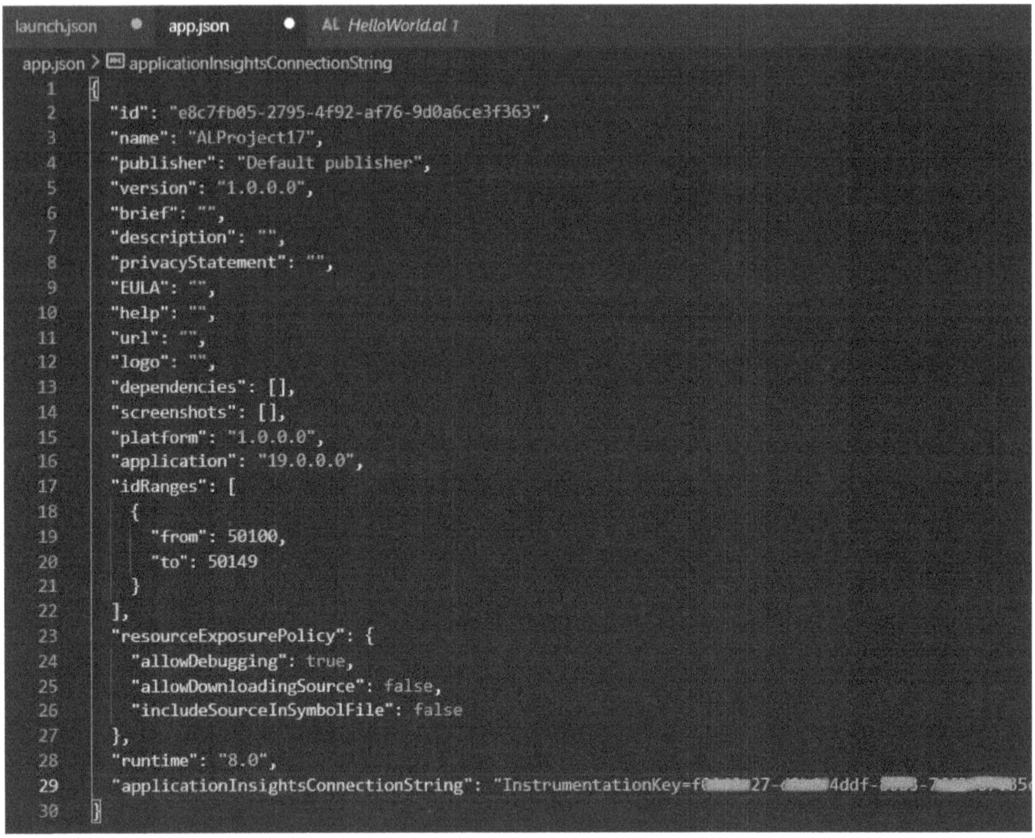

```
launch.json    ●    app.json    ●    AL  HelloWorld.al 1

app.json > 🔤 applicationInsightsConnectionString
  1   {
  2       "id": "e8c7fb05-2795-4f92-af76-9d0a6ce3f363",
  3       "name": "ALProject17",
  4       "publisher": "Default publisher",
  5       "version": "1.0.0.0",
  6       "brief": "",
  7       "description": "",
  8       "privacyStatement": "",
  9       "EULA": "",
 10       "help": "",
 11       "url": "",
 12       "logo": "",
 13       "dependencies": [],
 14       "screenshots": [],
 15       "platform": "1.0.0.0",
 16       "application": "19.0.0.0",
 17       "idRanges": [
 18           {
 19               "from": 50100,
 20               "to": 50149
 21           }
 22       ],
 23       "resourceExposurePolicy": {
 24           "allowDebugging": true,
 25           "allowDownloadingSource": false,
 26           "includeSourceInSymbolFile": false
 27       },
 28       "runtime": "8.0",
 29       "applicationInsightsConnectionString": "InstrumentationKey=f0    27-    4ddf-    3-7    85
 30   }
```

Figure 4.9 – Application Insights for extension

> **Important Note**
>
> You can use the same Application Insights to collect the telemetry from different apps and environments. More than that, it is not tenant-dependent. You could use Application Insights in your tenant and collect the telemetry from tenants of different customers. Remember that it could be costly.

Now everything is ready and we can start to work with the telemetry.

Azure Application Insights

Azure Application Insights retains your telemetry as a log record. The number of records per day is individual for any customer, but you will have a lot of data in any case. In this section, we will learn the main features of Application Insights and how to select and filter the data:

1. Open the **Azure portal** and run **Application Insights** there.

2. Click on **Logs** and close the **Query** window, which is appearing first. Application Insights uses the **Kusto Query Language** (**KQL**) for the read-only requests.

Figure 4.10 – Application Insights query window

3. Double-click on Application Insights – **pageViews** or just print **pageViews** in the query window. Choose a time range of 1 hour and you will see opened pages for the last hour. Press **Run**.

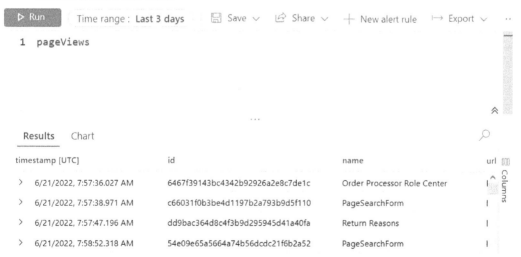

Figure 4.11 – Application Insights page views

4. If you type `traces` and run the query, you will see different operation types in your environment.

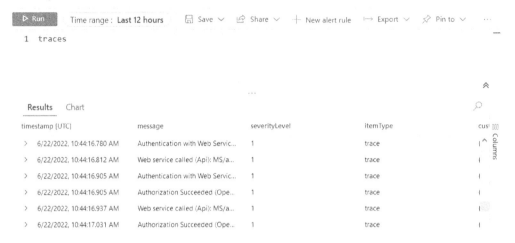

Figure 4.12 – Application Insights traces

5. If you want to export your telemetry, you can click on **Diagnostic Settings**, and then click on **+ Add diagnostic setting**.

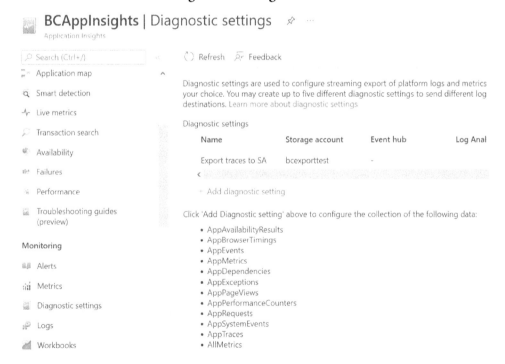

Figure 4.13 – Diagnostic settings

6.   Input the name of the diagnostic setting, and then select **AppPageViews** and/or **AppTraces**. If you left the retention policy as **0**, records will continue to be kept in Application Insights. If you choose some days in **Retention**, records after this number of days will be deleted after archiving. In **Destination details**, choose **Archive to a storage account**, and select a subscription and a storage account name. Click on **Save**. Now, your telemetry records will move to the storage account as a JSON file every hour.

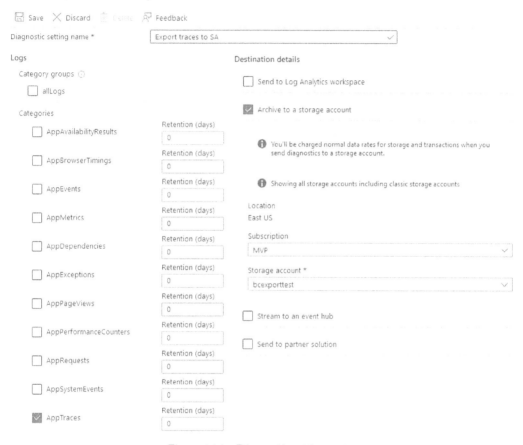

Figure 4.14 – Diagnostic settings setup

7. To get information about your costs in Application Insights, open the workspace link at the **Overview** tab and click **Usage and estimated costs**. You will see data usage over the course of the last month.

Usage Charts

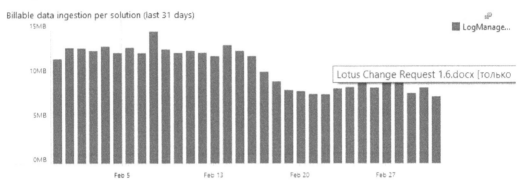

Figure 4.15 – Application Insights costs

8. Also, here you can change the **Data Retention** policy by clicking on it.

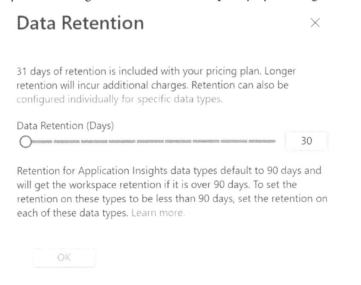

Figure 4.16 – Application Insights Data Retention

9.   With a **Daily cap** action, you can set daily data limits to reduce your costs.

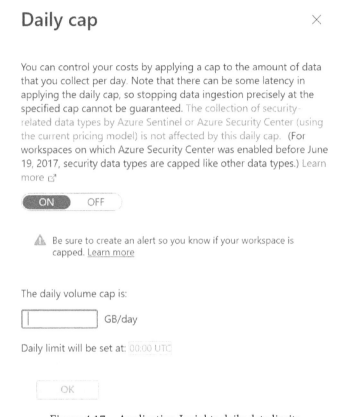

Figure 4.17 – Application Insights daily data limits

Now we can move on to the next section and learn how to analyze the telemetry.

Analyzing telemetry

In this section, we will learn how to find the requisite events with KQL and create telemetry alerts.

Let's open **Application Insights** in the Azure portal, select **Logs**, and open the **query** window. Start printing and just select variants from the drop-down menu.

First of all, we can check our logs for errors. **Severity levels** in Application Insights have different values than in the Admin portal.

Application Insights Severity Level	Description
0	Verbose
1	Information
2	Warning
3	Error
4	Critical

To find the errors, we need to select traces according to the third severity level. Therefore, print `traces` and press *Enter*. The new line will start with the | symbol. Print `where severityLevel == 3`. Choose the time interval that you want to analyze. You should get something like this:

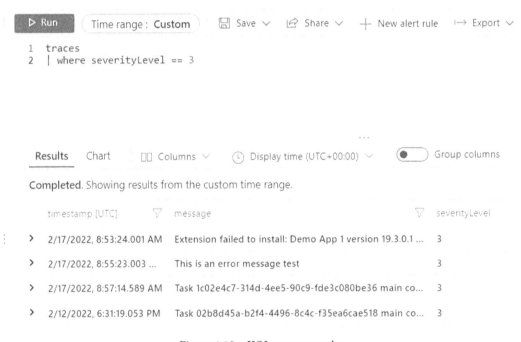

Figure 4.18 – KQL query sample

Expand the record and check for the **customDimensions** element. Here you can see many important values, such as **TenantId** (if you collect the telemetry for different tenants), **environmentName**, and **companyName**.

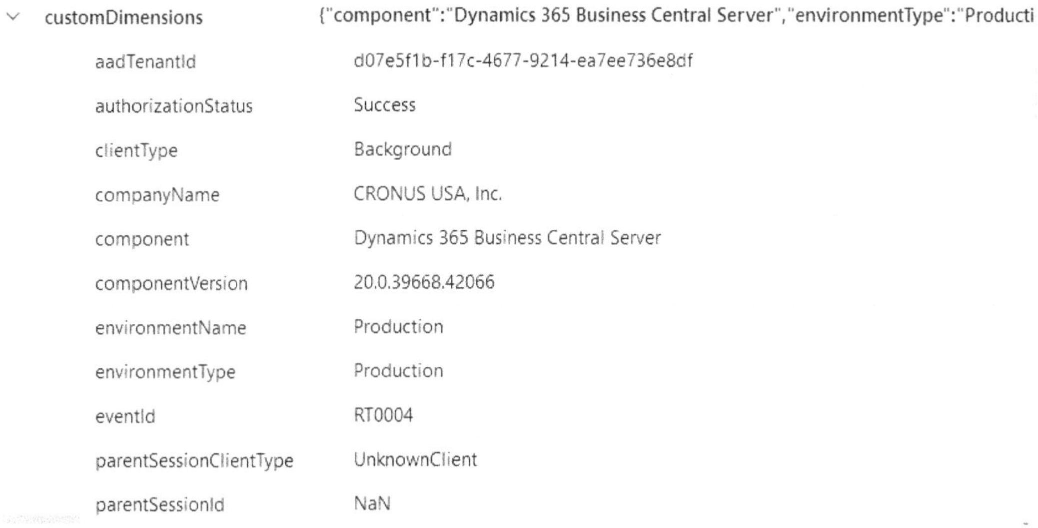

⌄	customDimensions	{"component":"Dynamics 365 Business Central Server","environmentType":"Producti
	aadTenantId	d07e5f1b-f17c-4677-9214-ea7ee736e8df
	authorizationStatus	Success
	clientType	Background
	companyName	CRONUS USA, Inc.
	component	Dynamics 365 Business Central Server
	componentVersion	20.0.39668.42066
	environmentName	Production
	environmentType	Production
	eventId	RT0004
	parentSessionClientType	UnknownClient
	parentSessionId	NaN

Figure 4.19 – customDimensions

To filter the telemetry additionally by tenant ID, you can apply the next request:

```
traces
| where severityLevel == 3 and customDimensions.aadTenantId ==
'Your Tenant ID'
```

Look at the **eventId** value under **customDimensions (LC0045)**. Each event type in the system has its own ID. You can use this parameter to filter the required event types. Here are some event IDs that may help you:

Event ID	Description
RT0001	Authorization failed (Pre Open Company)
RT0002	Authorization failed (Open Company)
RT0003	Authorization succeeded (Pre Open Company)
RT0004	Authorization succeeded (Open Company)
LC0001	Company created
LC0002	Company creation canceled
LC0003	Company creation failed
RT0012	Database lock timed out
RT0013	Database lock snapshot
AL0000CTV	Email sent successfully
AL0000CTP	Failed to send email
LC0010	Extension installed successfully
LC0011	Extension failed to install
LC0014	Extension published successfully
LC0015	Extension failed to publish
RT0010	Extension Update Failed: exception raised in extension
AL0000E25	Job queue entry started
AL0000E26	Job queue entry finished
RT0018	Operation exceeded time threshold (AL method)
RT0005	Operation exceeded time threshold (SQL query)
AL0000E2C	Permission set assigned to user
AL0000E2D	Permission set removed from user
RT0008	Web Services Call
RT0019	Web Service Called (Outgoing)
RT0020	Authentication with Web Service Key succeeded
RT0021	Authentication with Web Service Key failed

Now we can construct queries that are more interesting. For example, let's find unsuccessful Business Central API calls for the last hour. The query will be next:

```
traces
| where customDimensions.eventId == 'RT0008' and
customDimensions.httpStatusCode == 400
```

And the result is:

```
1  traces
2  |   where customDimensions.eventId == 'RT0008' and customDimensions.httpStatusCode == 400
```

...

Results Chart

timestamp [UTC]	message	severityLevel	itemType	customDimensions
> 6/22/2022, 8:31:29.260 AM	Web service called (Api): MS/ap...	1	trace	{"endpoint":"MS/api/r
> 6/22/2022, 8:31:29.401 AM	Web service called (Api): MS/ap...	1	trace	{"endpoint":"MS/api/r
> 6/22/2022, 8:31:29.417 AM	Web service called (Api): MS/ap...	1	trace	{"endpoint":"MS/api/r
> 6/22/2022, 8:46:29.476 AM	Web service called (Api): MS/ap...	1	trace	{"endpoint":"MS/api/r
> 6/22/2022, 8:46:29.603 AM	Web service called (Api): MS/ap...	1	trace	{"endpoint":"MS/api/r
> 6/22/2022, 8:46:30.077 AM	Web service called (Api): MS/ap...	1	trace	{"endpoint":"MS/api/r

Figure 4.20 – Failed API calls

You can use telemetry for your code analysis. If you run the next query:

```
traces
| where customDimensions.eventId == 'RT0018'
```

You will see the long-running AL operations and get the objects with a call stack where your code could not work optimally and requires a lot of time to execute.

If you assigned an Application Insights connection string to your app, you can monitor who installed, uninstalled, or updated your app using the following event IDs: LC0010, LC0016, and LC0022.

More information about telemetry event codes is available here:

```
https://docs.microsoft.com/en-us/dynamics365/business-central/
dev-itpro/administration/telemetry-event-ids
```

If you have created a new functionality and want to know whether users opened it, you can monitor **pageViews** instead of **traces**. This request will show you the number of your page views:

```
pageViews
| where  customDimensions.alObjectId == <<your page object ID>>
```

From **Business Central 2022 Release Wave 1**, you can assign a telemetry ID to the user card and find this ID in the telemetry traces to troubleshoot the user's problems. The ID is only included in traces that occur in the context of the user's session.

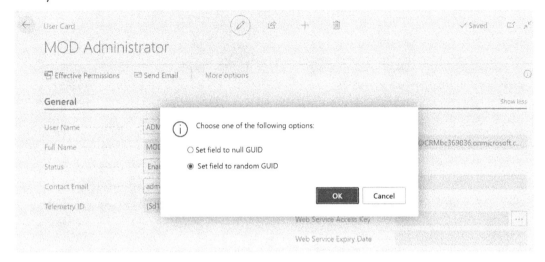

Figure 4.21 – User Telemetry ID

Custom events

By the way, you are not limited to pre-defined telemetry events. For your extension, you may log custom events using the standard `LogMessage` procedure. You can place them anywhere you need them. To test this, you can create a new project in the VS Code and perform the following steps:

1. Complete the `applicationInsightsConnectionString` value in the `app.json` file with the connection string to your Application Insights.

2. Open the default file with a Hello world message for the customers' list and replace the `OnOpenPage` trigger with the following code:

```
trigger OnOpenPage();
    var
        CustDimension: Dictionary of [Text, Text];
    begin
        CustDimension.Add('result', 'failed');
        CustDimension.Add('reason', 'critical error in
code');
        LogMessage('AB0001',
                   'This is an error message test',
```

```
                     Verbosity::Error,
                     DATACLASSIFICATION::SystemMetadata,
                     TelemetryScope::ExtensionPublisher,
                     CustDimension);
         end;
```

Here, `CustDimension` is a text dictionary where you can place all the requisite metrics, `AB0001` is a custom event ID (you can fill it as you wish, but remember not to use standard IDs), and `Verbosity` is the severity level.

3. Now, publish your extension to the environment and open the **Customers list** field if you changed `startupObjectId` and this list does not open automatically.

4. Then, go to Application Insights and check for the traces. You can add filters through `severityLevel = 3` or any other filters if you get a lot of data. Your event will be on the list.

Figure 4.22 – Custom telemetry event

You can also use a simplified variant of the `LogMessage` procedure without a text dictionary and with one or two dimensions like this:

```
trigger OnOpenPage();
    begin
        LogMessage('AB0001',
                    'This is an error message test',
                    Verbosity::Error,
                    DATACLASSIFICATION::SystemMetadata,
                    TelemetryScope::ExtensionPublisher,
                    'result', 'failed',
                    'reason', 'critical error in code');
    end;
```

As in the previous code sample, this will create a telemetry record with two custom dimensions – **result** and **reason**, but you do not need to declare a dictionary type variable.

Summary

Finally, you have come to the end of this chapter! This material was more complex than in the previous chapters. Telemetry setup and analysis require advanced skills and now you have them. You can collect the telemetry from the environment or from the extension. You know where to find your telemetry and how to select the necessary records and analyze them. More than that, you can create your own telemetry wherever you need it. Remember that with telemetry, you can learn about issues with your apps earlier than your customers.

In the next chapter, we will learn how to report issues and outages within your environment to Microsoft.

5
Reported Outages and Operations

As a tenant administrator, you could be the person who creates support tickets and reports about outages to Microsoft. In this chapter, we will look at the issue-handling process. In addition, we will learn how to check the environment operations log to see the possible reasons for the issue.

In this chapter, we are going to cover the following main topics:

- Reporting an outage
- Reporting an issue
- Managing your reported outages
- Environment operations analysis

By the end of this chapter, you will be able to create a support ticket, report a serious outage, and analyze operation log entries.

> **Important Note**
> Remember that the first line of support is your Microsoft Partner. Ask them first, before you create a support ticket with Microsoft.

Reporting an outage

We have two different situations when we might need Microsoft support:

1. The first is the production outage. It is a serious incident with two types, as shown in the following figure:

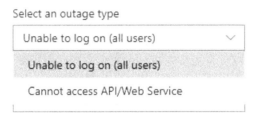

Figure 5.1 – Outage types

> **Important Note**
> Outage reporting is only available for the **Production** type environments.

2. The second one is the common issue. Here, we just need to create a support ticket. We will look into this in the next section.

Production outage

1. Let's start with outages. To report an outage, open your production environment card and choose **Support** and **Report Production Outage**.

Figure 5.2 – Report Production Outage

2. After that, you will see the report dialog where you should select the outage type, and fill in your first and last name, phone number, and email address. Press **Next** when you have done that:

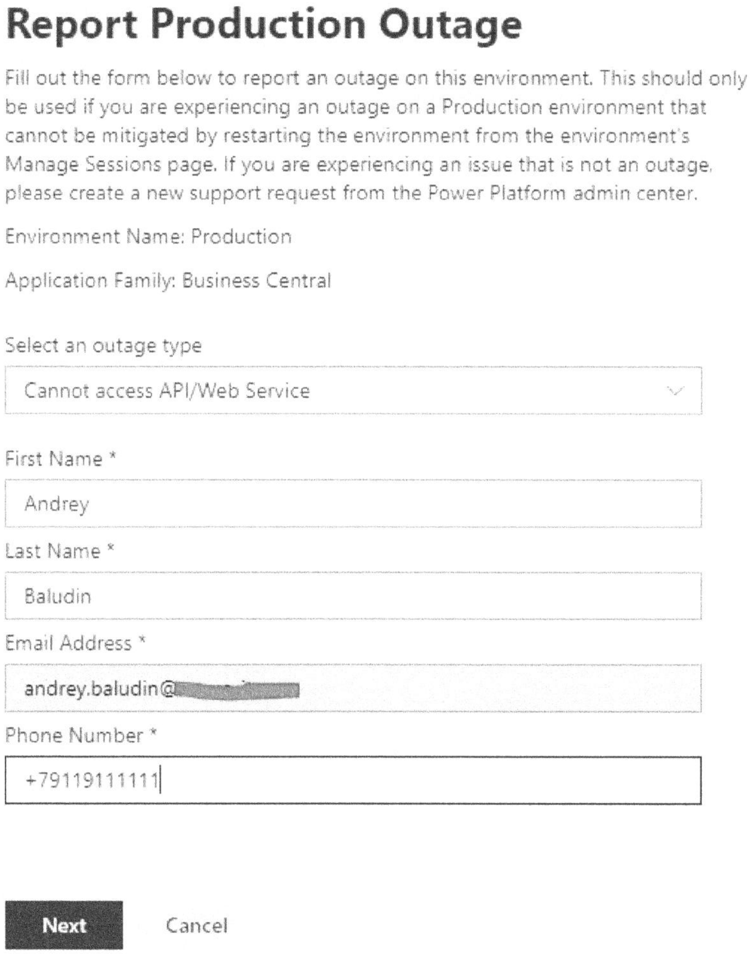

Report Production Outage

Fill out the form below to report an outage on this environment. This should only be used if you are experiencing an outage on a Production environment that cannot be mitigated by restarting the environment from the environment's Manage Sessions page. If you are experiencing an issue that is not an outage, please create a new support request from the Power Platform admin center.

Environment Name: Production

Application Family: Business Central

Select an outage type

Cannot access API/Web Service

First Name *

Andrey

Last Name *

Baludin

Email Address *

andrey.baludin@

Phone Number *

+79119111111

Next   Cancel

Figure 5.3 – Report Production Outage dialog

Depending on the selected outage type, you will see different windows.

For **Unable to log on**, you will need to do the following:

A. Confirm that you have tried to sign in with different browsers and list them.

B. Confirm that users cannot log in to any companies and if they can, list them.

C. Choose the outage start date and time.

D. Confirm the privacy statement.

E. Press **Report** to send the outage report.

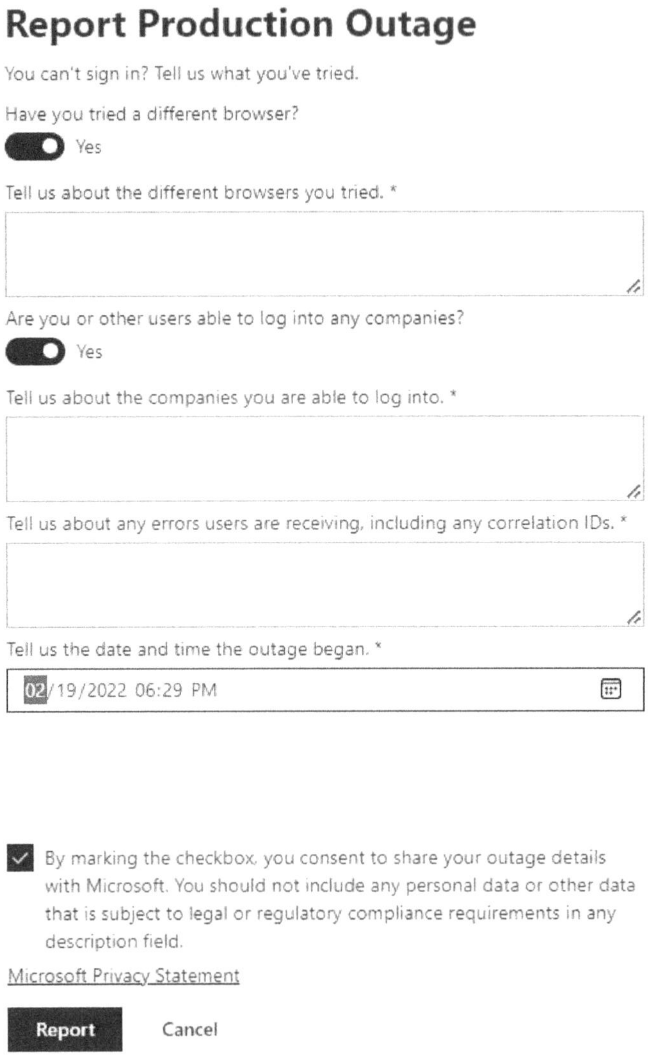

Report Production Outage

You can't sign in? Tell us what you've tried.

Have you tried a different browser?

[●○] Yes

Tell us about the different browsers you tried. *

[text area]

Are you or other users able to log into any companies?

[●○] Yes

Tell us about the companies you are able to log into. *

[text area]

Tell us about any errors users are receiving, including any correlation IDs. *

[text area]

Tell us the date and time the outage began. *

| 02/19/2022 06:29 PM | 📅 |

[✓] By marking the checkbox, you consent to share your outage details with Microsoft. You should not include any personal data or other data that is subject to legal or regulatory compliance requirements in any description field.

Microsoft Privacy Statement

[**Report**]   Cancel

Figure 5.4 – Sign-in outage

For **Cannot access API/Web Service**, you need to do the following:

A. List the non-working APIs and services.

B. Fill in the errors you receive.

C. Choose the outage start date and time.

D. Confirm the privacy statement.

E. Press **Report** to send the outage.

Report Production Outage

You can't connect to a web service or API? Tell us what you've tried.

Which web service/API is not working? *

Tell us about any errors users are receiving, including any correlation IDs. *

Are users still able to access Business Central with the client?

[●] Yes

Tell us the date and time the outage began. *

02/19/2022 06:06 PM

[✓] By marking the checkbox, you consent to share your outage details
with Microsoft. You should not include any personal data or other data
that is subject to legal or regulatory compliance requirements in any
description field.

Microsoft Privacy Statement

[Report]   Cancel

Figure 5.5 – API access outage

3. Microsoft will contact you by phone, email, or Teams. You can manage your
reported outages on the **Reported Outages** tab of the admin center:

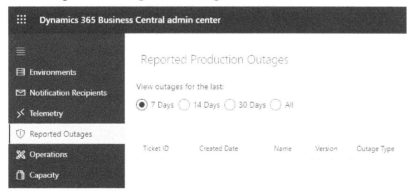

Figure 5.6 – Reported Outages

> **Note**
> Outages and breaking issues are free of charge.

Report issue

If neither your Microsoft Partner nor yourself as the delegated admin can resolve the customer's issue, you could create a support ticket with Microsoft.

Submitting a ticket through the admin portal

This is for a support ticket, which you are submitting on behalf of a customer:

1. First, you need to open your environment card and click on **Support** and **New Support Request**:

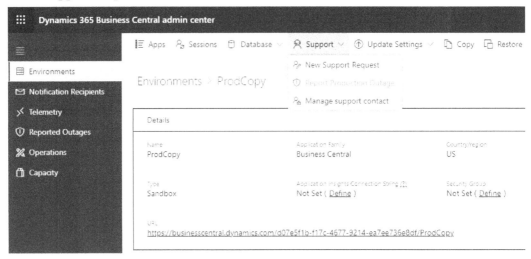

Figure 5.7 – New Support Request

2. After the click, **Power Platform admin center** will open. Here, you can see your previous support tickets. To create the new one, press **+ New support request**. The support center is also accessible by direct link: `https://admin.powerplatform.microsoft.com/support`.

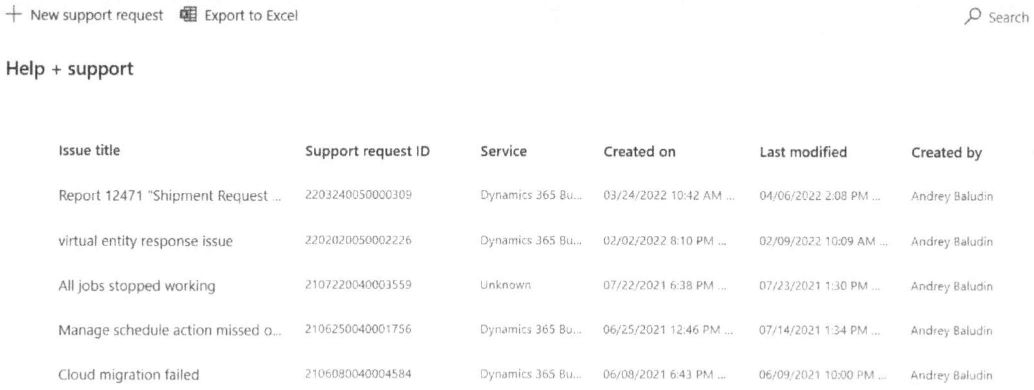

+ New support request   ⊞ Export to Excel   ○ Search

Help + support

Issue title	Support request ID	Service	Created on	Last modified	Created by
Report 12471 "Shipment Request ...	2203240050000309	Dynamics 365 Bu...	03/24/2022 10:42 AM ...	04/06/2022 2:08 PM ...	Andrey Baludin
virtual entity response issue	2202020050002226	Dynamics 365 Bu...	02/02/2022 8:10 PM ...	02/09/2022 10:09 AM ...	Andrey Baludin
All jobs stopped working	2107220040003559	Unknown	07/22/2021 6:38 PM ...	07/23/2021 1:30 PM ...	Andrey Baludin
Manage schedule action missed o...	2106250040001756	Dynamics 365 Bu...	06/25/2021 12:46 PM ...	07/14/2021 1:34 PM ...	Andrey Baludin
Cloud migration failed	2106080040004584	Dynamics 365 Bu...	06/08/2021 6:43 PM ...	06/09/2021 10:00 PM ...	Andrey Baludin

Figure 5.8 – Power Platform admin center support

3. The support request window will open. Here, you must provide details about your issue. Complete the following steps:

    A. Choose a product. In our situation, the most likely will be **Dynamics 265 Business Central**.

    B. Choose the **Problem type** and **Problem subtype** options from the respective lists.

    C. Choose the environment. If the Power Platform admin center belongs to the customer's tenant, you can choose an environment from the list. If you are the delegated admin and work in your partner's admin center, you must provide a link to the environment and choose the environment type.

D. The next fields depend on **Problem type**. Fill them out to clarify the issue, then click **See solutions**:

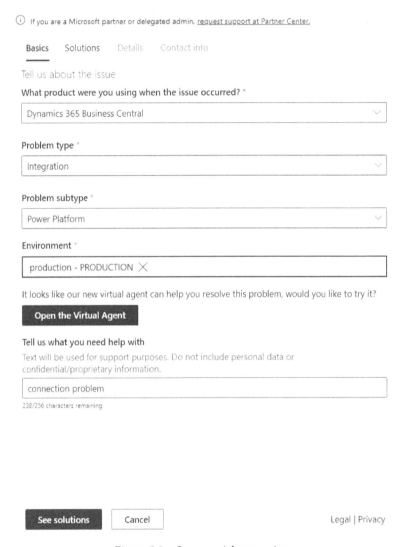

Figure 5.9 – Support ticket creation

4. Then, you will see the links to possible solutions. Please check them before continuing, as the solution is probably there. If you cannot find any helpful information, click on **Next**:

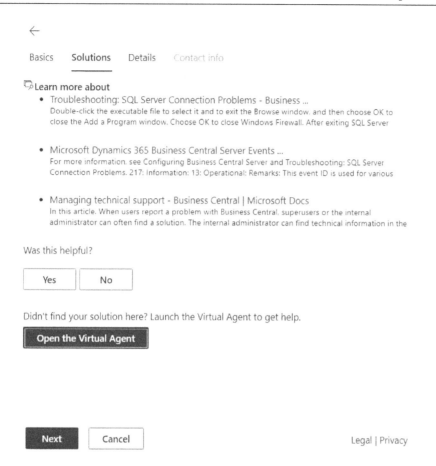

Figure 5.10 – Support ticket solutions

If you clicked **Next** as the provided links did not help you to solve the issue, follow these steps:

1. Choose your support plan and the level of severity (evaluate the real impact of the issue on your business). Subscription support is already included when you buy a Business Central license and you can get support for most critical cases free of charge. Professional direct support is available for an extra charge.

2. Fill in **Issue title** and **Issue description**. Choose whether the issue is related to a recent service change.

3. Attach files if needed.

4. Press **Next**:

Figure 5.11 – Support ticket details

For the last step, follow these steps:

1. Choose your country.
2. Fill in your email address and the email addresses of people you want to have copied in.
3. Fill in your phone number and preferable method of contact.
4. Press **Submit**.

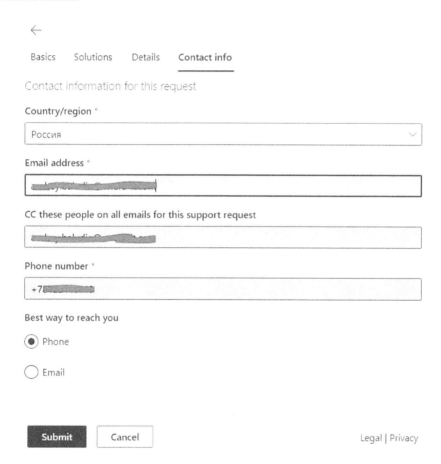

Figure 5.12 – Support ticket contact information

When this is all done, you will get an email saying that your request is registered. After that, a support specialist will contact you depending on the severity level. The response time for critical issues (severity A) is 1 hour. Other levels are dependent on your support plan.

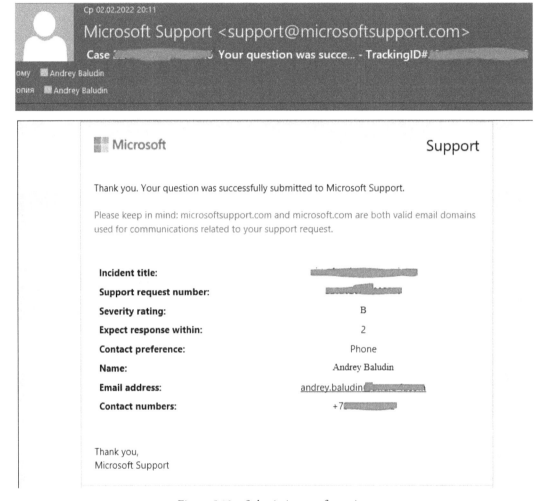

Figure 5.13 – Submission confirmation

Now, you are able to handle creating support tickets and we can now move to the next section, environment operations.

Environment operations analysis

To see the environment operations, open the admin center and click on the **Operations** tab. Here, you will see the significant operations on your environments. You can choose single environment operations or look through all environments. In addition, you can choose default periods – 7, 14, or 30 days, or a custom time interval.

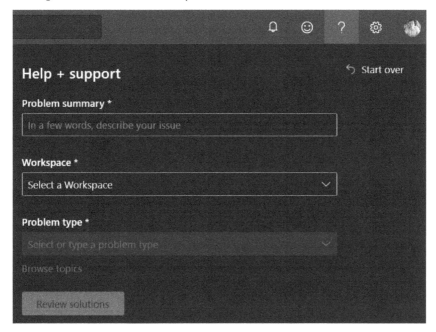

Figure 5.14 – Environment Operations

Possible operation types that go to the operations log are as follows:

- Environment creation
- Environment rename
- Environment delete
- Environment properties modification (for example, changing of update window, security group, or Application Insights connection string)
- Environment update (major and minor)
- Environment restore
- Environment copy

- Environment restart
- App install
- App update

By clicking on the operation status, **Complete** or **Failed**, you can see operation details, which will definitely be helpful for you.

Summary

You now have important knowledge – if you and your partner cannot resolve an issue, you can create a support ticket with Microsoft. If you and your colleagues cannot sign in to the environments (go back to *Chapter 1*, if you forgot how to construct the admin center direct link) or your APIs are not accessible, you can report an outage. You now know where you can create tickets and how to see the results. In addition, you know where to check operations in your environments. Now, we can move to the next chapter and learn how to manage your capacity.

6

Tenant Capacity Management

Compared with on-premises, SaaS environments have some limitations, and capacity is one of them. When you decide to implement and use Dynamics 365 Business Central, you must evaluate your business and the operations that you perform because your cloud storage is limited. If you are a **big customer** (we'll discuss what this is later in the chapter) and choose SaaS, you could be out of capacity after some time and have to buy capacity add-ons with extra costs. Another way is to set up retention policies and delete the old data or compress it. All of these processes will be covered in this chapter.

We are going to cover the following main topics:

- Storage capacity usage
- Environments quota usage
- Storage usage by environment
- Some practical advice on how to keep your data within limits
- Big customers

After this chapter, you will be able to manage your cloud storage and predict a lack of capacity. In addition, you will be able to create retention policies and evaluate your business processes.

Storage capacity usage

As this is a very important part of the administration process, Microsoft put **Capacity** management into a separate tab of **Admin Center**. You can find it in the last menu tab.

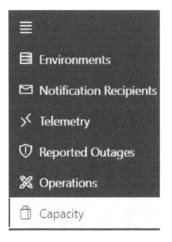

Figure 6.1 – Capacity tab

The first thing that you see is the **Storage capacity usage** tab. It shows you the total storage amount that you have for *all* of your environments, used and available space. You can look just at these numbers and understand whether everything is OK with your storage or you need to clear the space.

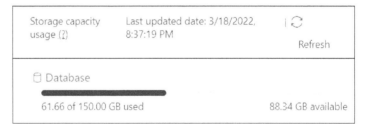

Figure 6.2 – Storage capacity usage

This tab also shows you **Last updated date and time** because the calculation of used storage takes time. If you press the **Refresh** button, you will be notified that calculation can take a long time. In fact, it performs in the background and you need to reopen the **Capacity** tab later to see the updated numbers.

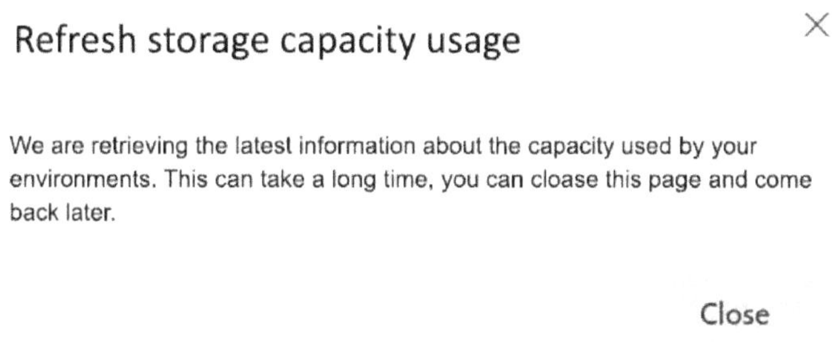

Figure 6.3 – Refresh storage capacity usage

> **Important Note**
> You can click on the **?** sign near the **Storage capacity usage** action to get information about capacity from Microsoft Docs: `https://docs.microsoft.com/en-us/dynamics365/business-central/dev-itpro/administration/tenant-admin-center-capacity`.

Storage capacity by source

On the next tab, you can see detailed capacity numbers. As we learned from *Chapter 1*, total capacity depends on the following parameters:

- Default capacity (80 GB)
- Type and quantity of user licenses (1, 2, or 3 GB per each)
- Additional capacity (environments quantity, purchased capacity add-ons)

From the user's perspective, it looks like this:

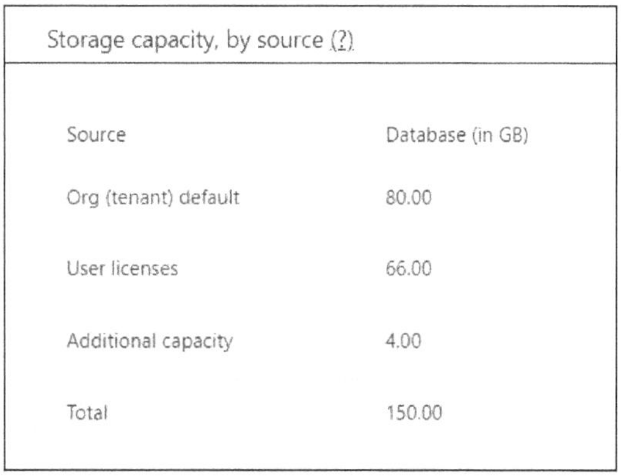

Storage capacity, by source (?)

Source	Database (in GB)
Org (tenant) default	80.00
User licenses	66.00
Additional capacity	4.00
Total	150.00

Figure 6.4 – Storage capacity by source

So, we can understand that from Microsoft's point of view, 80 GB is enough for small and medium companies' business operations, and the company could regulate this volume with user licenses. If a company has 25 essential users, then it has 50 GB of additional storage. Seems logical until you want to create an extra production environment.

Imagine that you open a branch office somewhere in a different country. For this office, you want to deploy Business Central with another localization. When you ask your partner to purchase one more production environment, you will get a surprise – your capacity will increase only by 4 GB – not the extra default 80 GB as some might think. Now your total capacity is shared between two production environments, and if 130 GB (80 GB default + 50 GB license) was enough for one environment, it is not enough for two.

More than that, imagine that you need to create a sandbox as a copy of the production environment. It will take just the same volume as production and you could be out of capacity. And what if you need copies of each production environment? The solution could be the usage of locally restored backups of the production environments, but it will require additional skills and actions. Remember these tricky moments when you are planning your architecture.

Environments quota usage

Under **Storage Capacity usage**, you can see the next tab – **Environments quota usage**.

Figure 6.5 – Environment quota usage

Just simple numbers. By default, you have one production environment and you can create up to three sandboxes. If you buy an additional production environment through your partner, you will also get three extra sandboxes. As you can see in *Figure 6.5*, this tenant has one default + one additional production environment and six available sandboxes. The customer has created five sandboxes and could create one more.

Storage usage by environment

At the end of the **Capacity** tab, you can find a very important thing for storage analysis – **Storage usage by environment**. It allows you to see the total environment's database usage and details about the environment's tables. You can see which tables are the biggest and think about this data: do you really need this data, or can you archive or even delete it?

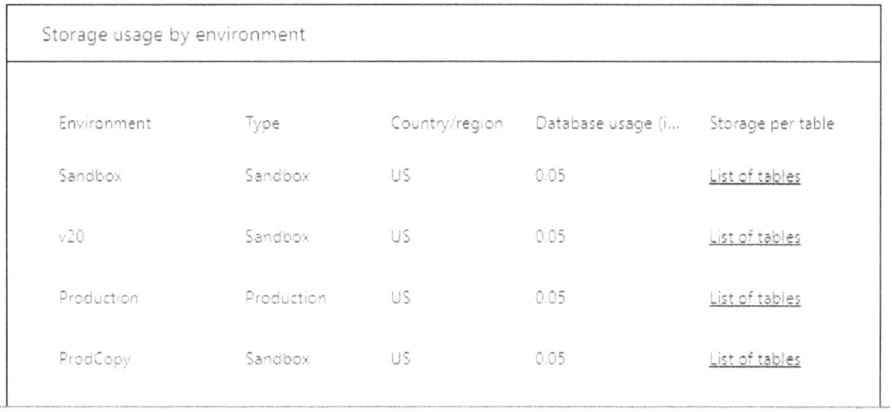

Storage usage by environment				
Environment	Type	Country/region	Database usage (i…	Storage per table
Sandbox	Sandbox	US	0.05	List of tables
v20	Sandbox	US	0.05	List of tables
Production	Production	US	0.05	List of tables
ProdCopy	Sandbox	US	0.05	List of tables

Figure 6.6 – Storage usage by environment

The **Storage usage by environment** tab has a simple structure. Just the most important information about your environments is included:

- **Name**
- **Type**
- **Country/region**

The most interesting parts are these:

- **Database usage**
- **Storage per table**

If you click on the **List of tables** link at the end of each environment's line, it will open a **Table Information** page in the selected Business Central environment. You will also be able to open it directly from the web client and not open the Admin Center.

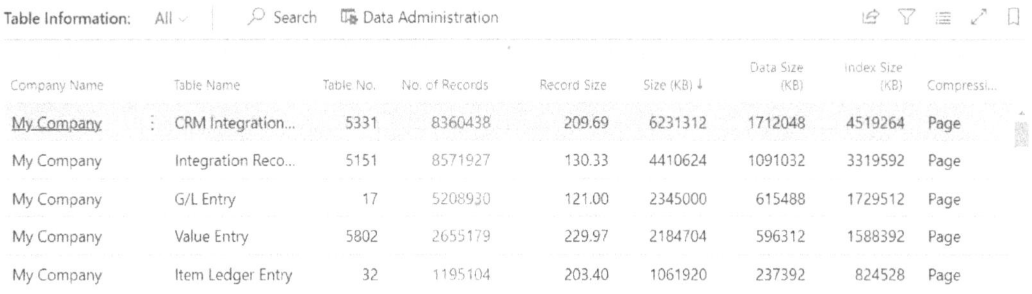
Table Information: All | Search | Data Administration

Company Name	Table Name	Table No.	No. of Records	Record Size	Size (KB) ↓	Data Size (KB)	Index Size (KB)	Compressi...
My Company	CRM Integration...	5331	8360438	209.69	6231312	1712048	4519264	Page
My Company	Integration Reco...	5151	8571927	130.33	4410624	1091032	3319592	Page
My Company	G/L Entry	17	5208930	121.00	2345000	615488	1729512	Page
My Company	Value Entry	5802	2655179	229.97	2184704	596312	1588392	Page
My Company	Item Ledger Entry	32	1195104	203.40	1061920	237392	824528	Page

Figure 6.7 – Storage usage by environment

This is the structure of your environment. At the top of the list, you can see the biggest tables you have. The most important details here are as follows:

- **Company Name**
- **Table Name** and **Table No.**
- **No. of Records**: number of table entries
- **Size (KB)**: table data size in KB

Using this, you can evaluate your business processes and understand which of them generates the main volume of data. In *Figure 6.7*, you can see that a lot of storage requires environment integration with **Dataverse** (through the **CRM integration record** table). Information about synchronized records is stored in the Business Central. Entry tables such as **General Ledger**, **Value**, and **Item Ledger** are also at the top of the list and this is absolutely expected.

Now you know how to get information about your data, so let's move on and learn some practical advice on how to reduce the data volume in your environment.

Some practical advice on how to keep your data within limits

In Business Central, you have a special tool to manage your data – **Data Administration**. You can open this page by clicking on the **Data Administration** action on the **Table Information** page or just open it directly from **Tell me**.

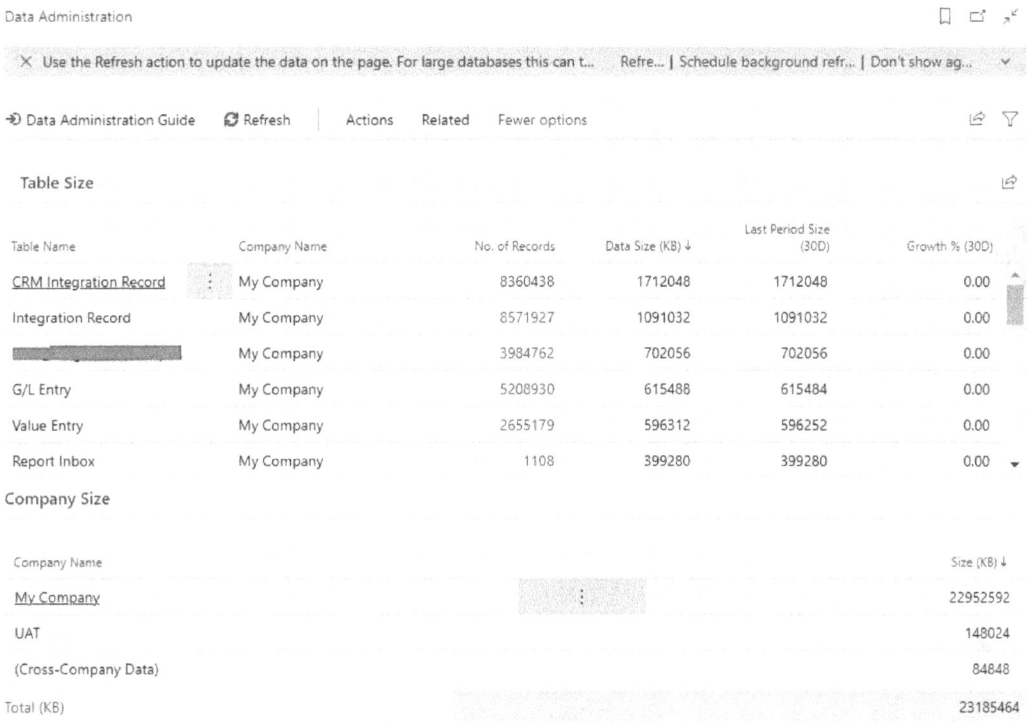

Data Administration

X Use the Refresh action to update the data on the page. For large databases this can t...   Refre... | Schedule background refr... | Don't show ag...   ∨

→Đ Data Administration Guide   ⟳ Refresh | Actions   Related   Fewer options

Table Size

Table Name	Company Name	No. of Records	Data Size (KB) ↓	Last Period Size (30D)	Growth % (30D)
CRM Integration Record	My Company	8360438	1712048	1712048	0.00
Integration Record	My Company	8571927	1091032	1091032	0.00
	My Company	3984762	702056	702056	0.00
G/L Entry	My Company	5208930	615488	615484	0.00
Value Entry	My Company	2655179	596312	596252	0.00
Report Inbox	My Company	1108	399280	399280	0.00

Company Size

Company Name	Size (KB) ↓
My Company	22952592
UAT	148024
(Cross-Company Data)	84848
Total (KB)	23185464

Figure 6.8 – The Data Administration tool

After the first opening, it will be empty and you need to click on the **Refresh** action to update the data. It takes some time and runs in the background. For frequent usage of **Data Administration**, click on **Actions** and then **Schedule Background Refresh**.

Data Administration

→Đ Data Administration Guide   ⟳ Refresh | Actions   Related   Fewer options

⊞ Schedule Background Refresh   Data Cleanup ∨   Date Compression ∨

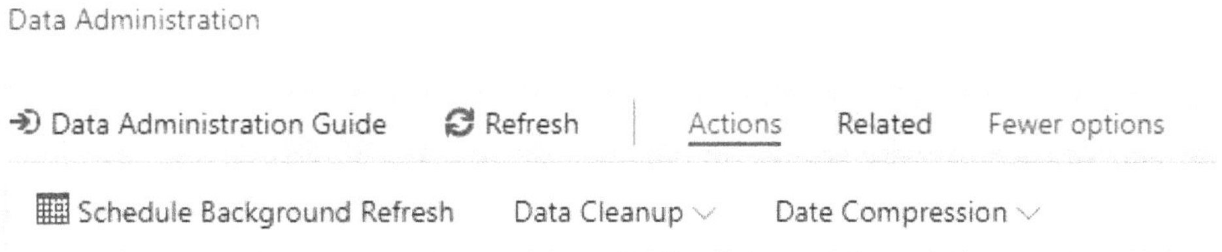

Figure 6.9 – Schedule background refresh

You will get a confirmation message:

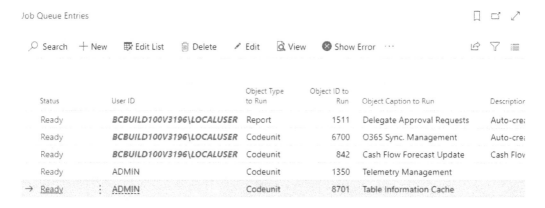

> (i) A job queue entry that runs daily to refresh the table information
> cache was created.

Figure 6.10 – Scheduled task creation confirmation

The new **Job Queue Category** and **Entry** will be created for scheduled information updates. You need not wait for data updates anymore while opening the **Data Administration** page. Open the **Job Queue Entries** list and ensure that the job has been created:

Job Queue Entries

🔍 Search   + New   📋 Edit List   🗑 Delete   ✏ Edit   🔍 View   ⊗ Show Error   ⋯

Status	User ID	Object Type to Run	Object ID to Run	Object Caption to Run	Description
Ready	*BCBUILD100V3196\LOCALUSER*	Report	1511	Delegate Approval Requests	Auto-crea
Ready	*BCBUILD100V3196\LOCALUSER*	Codeunit	6700	O365 Sync. Management	Auto-crea
Ready	*BCBUILD100V3196\LOCALUSER*	Codeunit	842	Cash Flow Forecast Update	Cash Flov
Ready	ADMIN	Codeunit	1350	Telemetry Management	
→ Ready	ADMIN	Codeunit	8701	Table Information Cache	

Figure 6.11 – Table Information Cache job

You have three options here to reduce the volume of your data:

- Clean up unimportant data.
- Compress the data.
- Set up retention policies.

Let's learn about each of these options in detail.

Cleaning up data

Data cleanup is used for the batch deletion of needed documents filtered by some parameters you choose.

Open the **Data Administration** page of Business Central and choose **Actions** and then **Data Cleanup**. By default, Microsoft already created for you a set of reports, which could help you to delete unused data. All of them you can run from this menu. I have summarized all available reports in the following table:

Cleanup type	Records to clean up
Logs	Change Log Entries
Document Archives	Expired Sales Quotes
	Sales Quote Versions
	Blanket Sales Orders Versions
	Sales Order Versions
	Purchase Quote Versions
	Purchase Order Versions
	Blanket Purchase Order Versions
Invoiced Documents	Blanket Sales Orders
	Sales Orders
	Sales Return Orders
	Blanket Purchase Orders
	Purchase Orders
	Purchase Return Orders
	Registered Warehouse Documents
Marketing	Campaign Entries
	Logged Segments
	Opportunities
	Tasks
	Interaction Log Entries
Cost Accounting	Cost Budget Entries
	Cost Entries
	Old Cost Entries
Miscellaneous	Physical Inventory Ledger

You can run a cleanup report of the needed documents, set up filters, and press **OK**. After that, it deletes the filtered documents. This way, you resolve two issues:

- You free your storage.

- You remove all the old and unused documents from the list.

Date compression

Entry tables are usually the biggest tables in any Business Central database, for example, General Ledger Entry tables, Value Entry tables, Item Ledger Entry tables, and so on. If your data is older than 5 years old, you can compress it. In fact, data compression is a combination of several table records into one new record. To perform it, your records must adhere to the following constraints:

- Be older than 5 years

- Be closed (**Field Open** must have the value **No**)

- Relate to the closed fiscal year

To compress your data, open the **Data Administration** page and press **Actions** and then **Date Compression**.

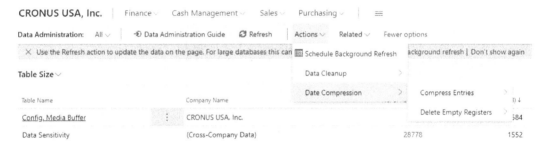

Figure 6.12 – Date Compression

You can select the date range for the records to be compressed and choose the fields that you want to have saved content. For example, this is a report to compress General Ledger Entries:

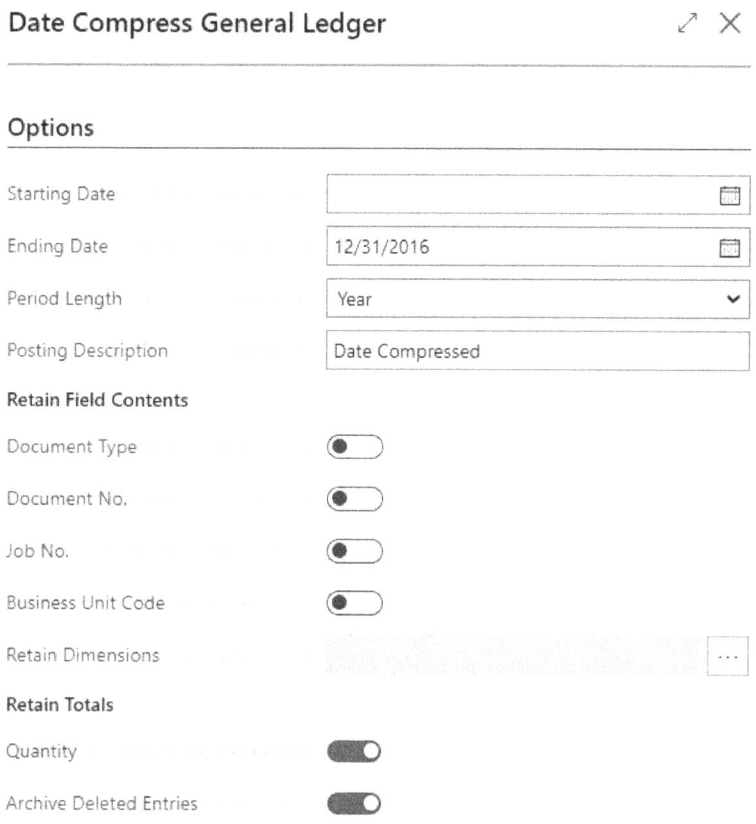

Date Compress General Ledger

Options

Starting Date	
Ending Date	12/31/2016
Period Length	Year
Posting Description	Date Compressed

Retain Field Contents

Document Type	
Document No.	
Job No.	
Business Unit Code	
Retain Dimensions	...

Retain Totals

Quantity	
Archive Deleted Entries	

Figure 6.13 – Date Compress General Ledger

You can compress the following data:

- General Ledger Entries
- Tax Entries
- Bank Account Ledger Entries
- General Ledger Budget Entries
- Customer Ledger Entries
- Vendor Ledger Entries
- Resource Ledger Entries
- Fixed Assets Ledger Entries
- Maintenance Ledger Entries

- Insurance Ledger Entries
- Warehouse Entries
- Item Budget Entries

Also, from the **Date compression** menu, you could select **Delete Empty Registers**.

> **Important Note**
> You must update all your analysis views before running date compression.
> For this, open the **Analysis Views** page and click the **Update** button.

Let's look at the compression example. We have a pack of entries in the year 2016, which we want to compress:

Entry No. ↑	G/L Account No.	Posting Date ▼	Document Type	Document No.	Description	Bal. Account No.	Amount
4296 ⋮ 10200		8/1/2016		123	Saving account	10300	100.00
4297	10300	8/1/2016		123	Saving account	10200	-100.00
4298	10200	8/9/2016		123	Saving account	10300	200.00
4299	10300	8/9/2016		123	Saving account	10200	-200.00
4300	10200	8/25/2016		123	Saving account	10300	200.00
4301	10300	8/25/2016		123	Saving account	10200	-200.00

Figure 6.14 – G/L entries before compression

After the compression job runs, these records will be deleted, and new records will be created:

Entry No. ↑	G/L Account No.	Posting Date ▼	Document Type	Document No.	Description	Bal. Account No.	Amount
4302 ⋮ 10200		1/1/2016			Date Compressed		500.00
4303	10300	1/1/2016			Date Compressed		-500.00

Figure 6.15 – G/L entries after compression

For regular data compression, you can run these reports as scheduled jobs.

> **Important Note**
> Data compression locks the tables, and users will not be able to perform some operations. Do not run data compression during business hours.

You could compress the data for several tables using **Data Administration Guide** from the **Data Administration** page. It is a wizard page that walks you through all the steps where you can choose a set of tables to compress.

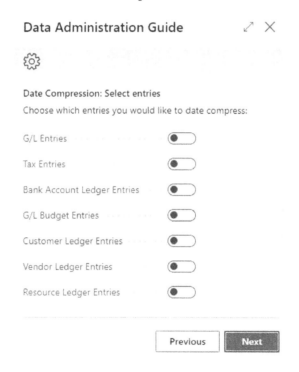

Figure 6.16 – Data Administration Guide

Data retention policies

If you do not need some old data at all, you can set up automatic data deletion with retention policies. This can be useful to clean up old documents and buffer tables.

You can open the **Retention Policies** page directly from the **Data Administration** page by clicking on **Related** and then **Retention Policies**. Alternatively, you can open it from the **Tell Me** option, printing **Retention Policies** there.

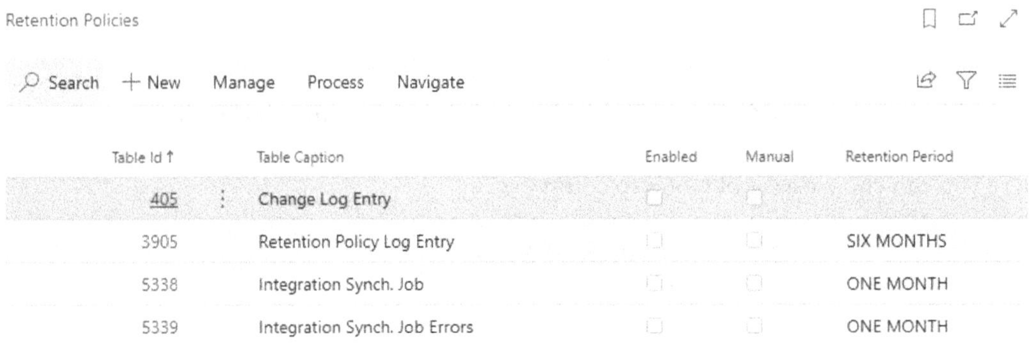

Table Id ↑	Table Caption	Enabled	Manual	Retention Period
405	Change Log Entry	☐	☐	
3905	Retention Policy Log Entry	☐	☐	SIX MONTHS
5338	Integration Synch. Job	☐	☐	ONE MONTH
5339	Integration Synch. Job Errors	☐	☐	ONE MONTH

Figure 6.17 – Retention Policies

You already have four policies but they are not enabled. By default, you are able to use the retention policy for the following tables:

- Change Log Entries
- Retention Policy Log Entries
- Integration Synch. Job
- Integration Synch. Job Errors
- Job Queue Log Entry
- Report Inbox
- Workflow Step Instance Archive
- Sales Header Archive
- Purchase Header Archive
- Email Outbox
- Sent Email

To enable a policy, open the policy's card and click on **Enabled**.

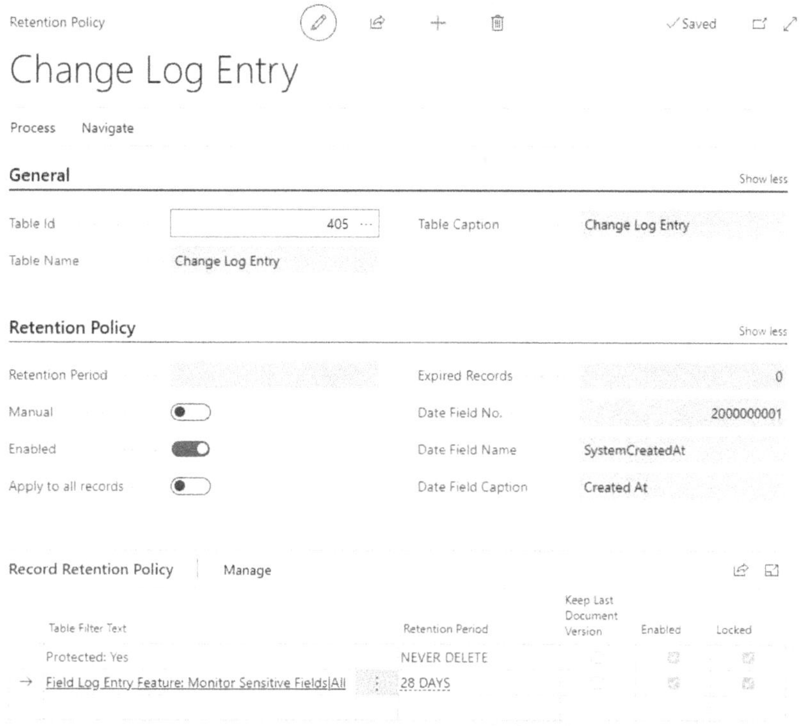

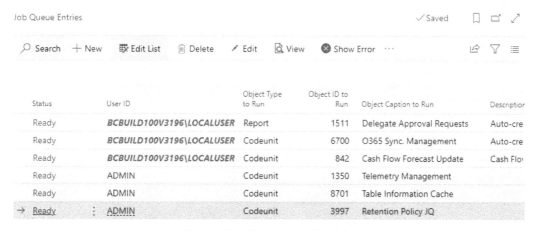

Figure 6.18 – Retention policy card

After that, it creates a scheduled job to run policies. This job applies all policies that are not marked **Manual**. You can run the policy manually by clicking on **Process** and **Apply Manually**.

Figure 6.19 – Retention policy job

If the policy is not marked **Apply to all records**, you can set a different behavior for records. For example, you may set things so that you never delete certain records that you have filtered.

You can create retention policies for other tables, not only the ones listed at the beginning of the section. However, you need to write some code for this. By default, when you try to create a new policy and try to select a table, you will see only default tables. To add your table to this list, publish an extension with the next install `codeunit`:

```
codeunit 5000 "My retention policies"
{
    Subtype = Install;


    trigger OnInstallAppPerCompany()
    var
        RetenPolAllowedTables: Codeunit "Reten. Pol. Allowed
Tables";
    begin
        RetenPolAllowedTables.AddAllowedTable(Database::"CHOOSE
YOUR TABLE");
    end;
}
```

After publishing, you will see your table in the policy's table list.

You can read more about retention policy creation here:

`https://demiliani.com/2020/11/18/dynamics-365-business-central-creating-retention-policies-from-al/`

> **Important Note**
> Here's an extra tip to keep your database within limits – do not store files in the Business Central database. Keep them in an **Azure Storage account** or **OneDrive**.

I hope that this advice will help you to keep your data up to date and within limits. Now we can move to the last section and review **big customers**.

Big customers

At the beginning of this section, I want to say that the numbers mentioned ahead are not limits for Business Central SaaS. The environment could operate using these numbers, but from my point of view, if you reach them, it will be better to have on-premises Business Central.

I did not pull these numbers out of thin air – they were presented at a Microsoft session about large customers at the Directions EMEA 2021 conference. These numbers are true at least for the years 2021-2022 and could change in the future.

So, you are a big customer if your single environment meets one of the following conditions:

- You have **1,000** records in the Users table.
- You have **800,000** records in the Sales Invoice Header table.
- You have **17 million** records in the G/L Entry table.
- Your database is more than **400 GB**.
- You migrate more than **160 GB** from on-premises environments.
- Your users perform more than **82,000** interactions in the browser **per hour**.
- Your users perform more than **385,000** interactions in the browser **per day**.
- You have more than **5,700** web service calls **per minute**.
- You have more than **2.7 million** web service calls **per day**.
- You have **6,000** sales orders posted **per day**.
- You have **3,300** lines **per sales order**.

These numbers are quite impressive. Please compare them with those of your own business.

> **Note**
>
> I repeat: if you exceed these numbers, it does not mean that you cannot use Business Central SaaS. It just means that on-premises installation could be more suitable for you, at least because of costs and performance.

Summary

You have finished this chapter and with that, you now know your tenant capacity. You know how many environments you can have and how much storage you can have. You know how to keep your data within limits and how to delete and compress data. If you need to deploy Business Central, you can evaluate your environment and think about environment installation types. Now we can move on to learn how to automate administration processes using the Admin Portal APIs.

7
Admin Center APIs

If you have repeatable actions with Dynamics 365 Business Central environments, want to automate some tasks, or even create your own interface for the Admin Center, you can make use of the Admin Center **application programming interfaces** (**APIs**). You could perform almost all these actions without opening the Admin Center. Create and delete environments, schedule updates, and set up notifications—you could run all these operations with a few clicks by calling the APIs.

In this chapter, we are going to cover the following main topics:

- Setting up API authentication
- Calling APIs
- Application APIs
- Environment operations APIs
- Environment settings APIs
- Other APIs

By the end of this chapter, you will be able to set up API authentication; you will also know which APIs exist for the Admin portal and how to call them with PowerShell and Postman.

Setting up API authentication

For authentication, Admin Center APIs use **Open Authorization 2.0** (**OAuth 2.0**) with **Azure Active Directory** (**Azure AD**). You must create an Azure app registration and let it call your Admin Center API. The process might be a bit tricky initially, but after a bit of practice, it should get a lot easier. Follow these steps to set up API authentication:

1. Open the **Azure portal** and find **App registrations**. Click on **+ New registration** to begin the process, as illustrated in the following screenshot:

Figure 7.1 – New app registration

2. In the second step, fill in a **Name** value for the app—for example, **Business Central**. For the **Supported account types** section, choose who will use the API—accounts from only your tenant or somebody else. Then, choose **Public client/Native** as the **Redirect URIs** value and type `BusinessCentralWebServiceClient://` `auth` in the box next to it. Click on **Register** to continue. The process is illustrated in the following screenshot:

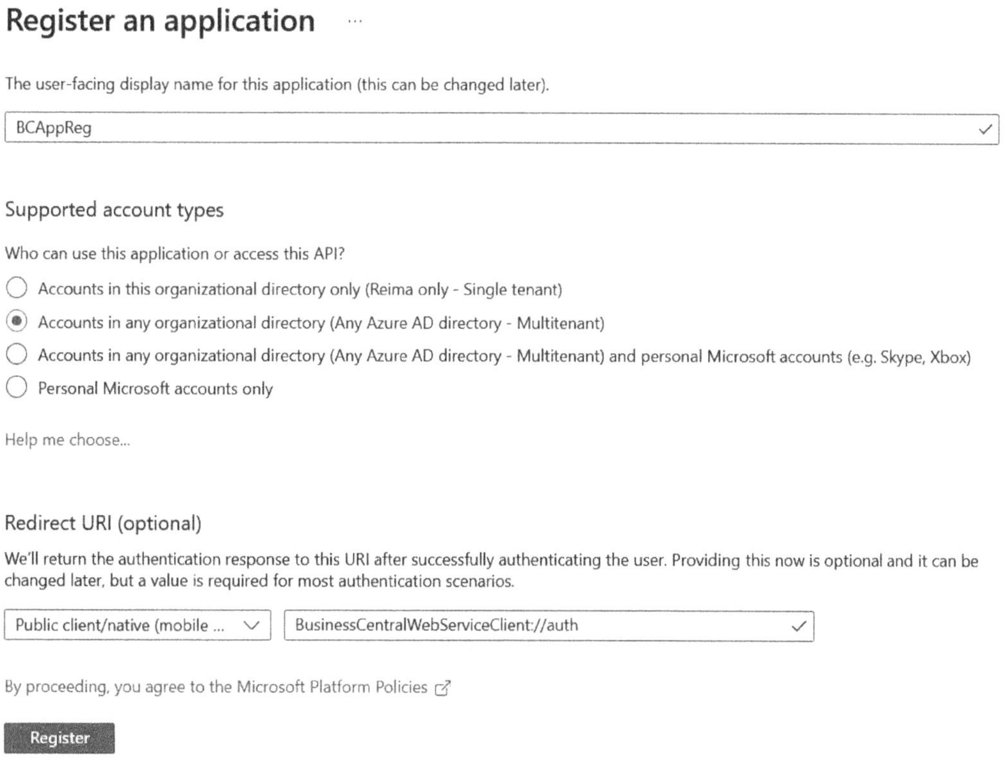

Register an application   ...

The user-facing display name for this application (this can be changed later).

> BCAppReg

Supported account types

Who can use this application or access this API?

○ Accounts in this organizational directory only (Reima only - Single tenant)

⦿ Accounts in any organizational directory (Any Azure AD directory - Multitenant)

○ Accounts in any organizational directory (Any Azure AD directory - Multitenant) and personal Microsoft accounts (e.g. Skype, Xbox)

○ Personal Microsoft accounts only

Help me choose...

Redirect URI (optional)

We'll return the authentication response to this URI after successfully authenticating the user. Providing this now is optional and it can be changed later, but a value is required for most authentication scenarios.

| Public client/native (mobile ... ∨ | BusinessCentralWebServiceClient://auth |

By proceeding, you agree to the Microsoft Platform Policies ⌕

Register

Figure 7.2 – Registering an application

3. After a few moments, your app registration will be created. Pay attention to important values such as **Application (client) ID** and **Directory (tenant) ID** as you will need these further. You can see these values in the following screenshot:

Figure 7.3 – App overview

4. Now, we need to provide permissions for your app registration. Click on the **API permissions** tab and then click **+ Add a permission**, as illustrated in the following screenshot:

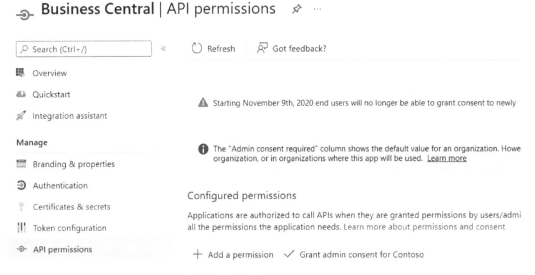

Figure 7.4 – Adding a permission

5. Choose **Dynamics 365 Business Central**, as shown here:

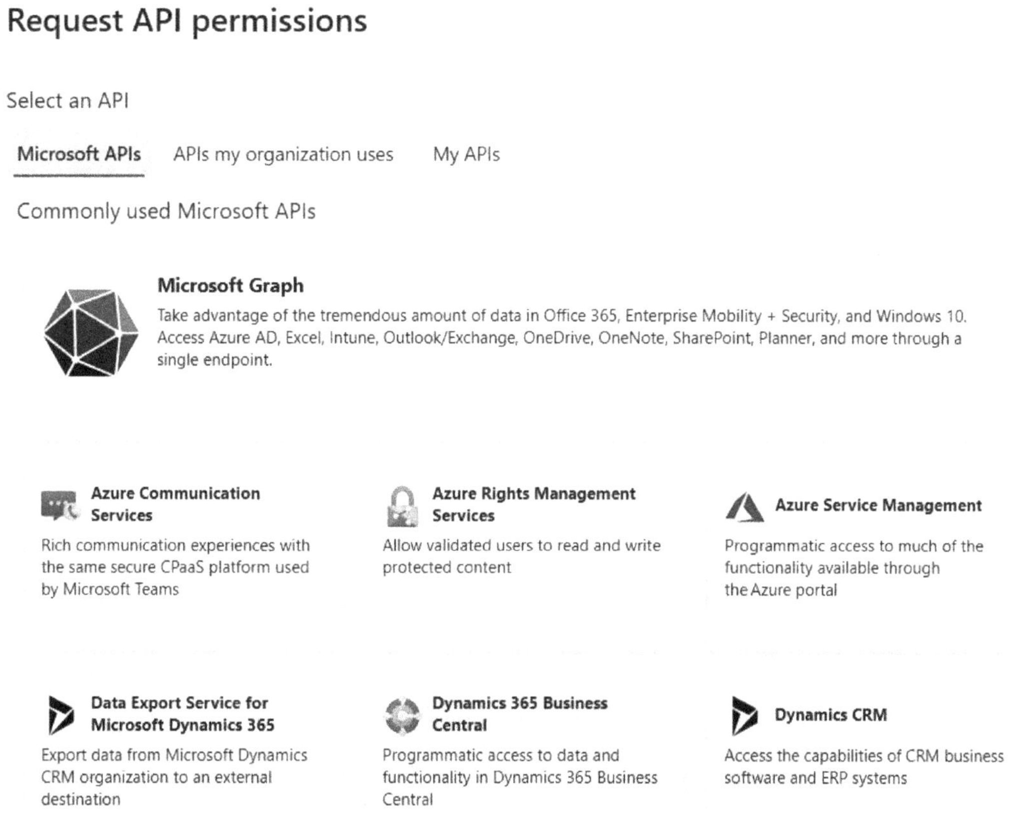

Request API permissions

Select an API

 **Microsoft APIs**   APIs my organization uses   My APIs

Commonly used Microsoft APIs

 Microsoft Graph
Take advantage of the tremendous amount of data in Office 365, Enterprise Mobility + Security, and Windows 10. Access Azure AD, Excel, Intune, Outlook/Exchange, OneDrive, OneNote, SharePoint, Planner, and more through a single endpoint.

Azure Communication Services
Rich communication experiences with the same secure CPaaS platform used by Microsoft Teams

Azure Rights Management Services
Allow validated users to read and write protected content

Azure Service Management
Programmatic access to much of the functionality available through the Azure portal

Data Export Service for Microsoft Dynamics 365
Export data from Microsoft Dynamics CRM organization to an external destination

Dynamics 365 Business Central
Programmatic access to data and functionality in Dynamics 365 Business Central

Dynamics CRM
Access the capabilities of CRM business software and ERP systems

Figure 7.5 – Requesting API permissions

6. Then, choose **Delegated permissions** and check both permissions listed. Following this, click **Add permissions** at the bottom of the screen, as illustrated in the following screenshot:

What type of permissions does your application require?

Delegated permissions	Application permissions
Your application needs to access the API as the signed-in user.	Your application runs as a background service or daemon without a signed-in user.

Select permissions   expand all

🔍 Start typing a permission to filter these results

ⓘ The "Admin consent required" column shows the default value for an organization. However, user consent can be customized per   ✕
permission, user, or app. This column may not reflect the value in your organization, or in organizations where this app will be
used. Learn more

Permission	Admin consent required
∨ Other permissions (1)	
☑ user_impersonation ⓘ Access as the signed-in user	No
∨ Financials (1)	
☑ Financials.ReadWrite.All ⓘ Access Dynamics 365 Business Central as the signed-in user	No

Figure 7.6 – Adding permissions

Your app registration is now ready, and we can try to get an **access token** and do something with the APIs.

Calling APIs

In this section, we will learn how to call Admin Center APIs. I suggest doing this in two different ways, as follows:

- From PowerShell, to show automation scenarios
- From Postman, to show a user-friendly interface

For example, you often update development sandboxes from a test or production sandbox, and you don't want to open the Admin portal each time and click through all those settings. You could create a PowerShell script and just run it each time you need to have a new environment.

Postman is one of the most popular **REpresentational State Transfer (REST)** clients. You can use it to test APIs, for single calls or demonstration purposes.

In this section, we will look into both environments.

PowerShell

1. Run **PowerShell** as an administrator. We need to install the `AzureAD` module and the `Microsoft.Identity.Client` package (if you haven't already done so). For this, run the next script:

    ```
    Install-Module AzureAD
    Install-Package -Name Microsoft.Identity.Client
    ```

2. Then, update the paths, like so:

    ```
    Add-Type -Path "C:\Program Files\WindowsPowerShell\
    Modules\Microsoft.Identity.Client\4.42.1\Microsoft.
    Identity.Client.dll"
    Add-Type -Path "C:\Program Files\WindowsPowerShell\
    Modules\AzureAD\2.0.2.140\Microsoft.IdentityModel.
    Clients.ActiveDirectory.dll"
    ```

    4.42.1 is the current version of the `Microsoft.Identity.Client` package and **2.0.2.140** is the current version of the `AzureAD` module. This may change in the future, so check the paths if you get an **unable to find the path** error.

3. Now, create variables with the values from the *Setting up API authentication* section. Use `$TenantId` as the **Directory (tenant) ID** value, `$AppId` as the **Application (client) ID** value, and `$RedirectUri` as the **Redirect URIs** value, as illustrated in the following code snippet:

    ```
    $TenantId = "d07e5f1b-f17c-4677-9214-ea7ee736e8df"
    $AppId = "810debd2-630d-4c3d-b645-bae2c8694d23"
    $RedirectUri = "BusinessCentralWebServiceClient://auth"
    ```

    With the following code, you can obtain an **access token**:

    ```
    $Client = [Microsoft.Identity.Client.
    PublicClientApplicationBuilder]::Create($AppId).
    WithAuthority("https://login.microsoftonline.
    com/$TenantId").WithRedirectUri($RedirectUri).Build()
    $Result = $Client.
    AcquireTokenInteractive([string[]]"https://api.
    businesscentral.dynamics.com/.default").ExecuteAsync().
    ```

```
GetAwaiter().GetResult()
$Token = $Result.AccessToken
```

If you run all the code from *step 3*, you will get a sign-in window where you must input your account name and password, as illustrated in the following screenshot:

Figure 7.7 – Sign-in window

After you click **Sign in**, you should see that your script is executed without errors, and the access token is kept in the $Token variable.

4. To check that everything works, let's get a list of environments from the **Admin portal**. You can find this at the following link: https://api. businesscentral.dynamics.com/admin/v2.11/applications/ environments.

    Or, you can call this from PowerShell using the following code:

    ```
    $response = Invoke-WebRequest `
        -Method Get `
        -Uri    https://api.businesscentral.dynamics.com/
    admin/v2.11/applications/environments `
        -Headers @{Authorization=("Bearer $Token")}
    Write-Host (ConvertTo-Json (ConvertFrom-Json $response.
    Content))
    ```

After you run the preceding code, you will see your list of environments, as follows:

```
$response = Invoke-WebRequest `
    -Method Get `
    -Uri    "https://api.businesscentral.dynamics.com/admin/v2.11/applications/environments" `
    -Headers @{Authorization=("Bearer $token")}
Write-Host (ConvertTo-Json (ConvertFrom-Json $response.Content))
{
    "value": [
        {
            "friendlyName":  "CRMbc369836-v20",
            "type":  "Sandbox",
            "name":  "v20-2",
            "countryCode":  "US",
            "applicationFamily":  "BusinessCentral",
            "aadTenantId":  "d07e5f1b-f17c-4677-9214-ea7ee736e8df",
            "applicationVersion":  "20.0.36751.0 (Preview)",
            "status":  "Active",
            "webClientLoginUrl":  "https://businesscentral.dynamics.com/d07e5f1b-f17c-4677-9214-ea7ee736e8df/v20-2",
            "webServiceUrl":  "https://api.businesscentral.dynamics.com/v2.0/d07e5f1b-f17c-4677-9214-ea7ee736e8df/v20-2",
            "locationName":  "East US",
            "platformVersion":  "20.0",
            "databaseSize":  "@{value=327155712.0; unit=Byte}",
            "ringName":  "Preview",
            "appInsightsKey":  ""
        },
        {
            "friendlyName":  "CRMbc369836-ProdCopy",
            "type":  "Sandbox",
            "name":  "ProdCopy",
            "countryCode":  "US",
            "applicationFamily":  "BusinessCentral",
            "aadTenantId":  "d07e5f1b-f17c-4677-9214-ea7ee736e8df",
            "applicationVersion":  "19.5.36567.36634",
            "status":  "Active",
            "webClientLoginUrl":  "https://businesscentral.dynamics.com/d07e5f1b-f17c-4677-9214-ea7ee736e8df/ProdCopy",
            "webServiceUrl":  "https://api.businesscentral.dynamics.com/v2.0/d07e5f1b-f17c-4677-9214-ea7ee736e8df/ProdCopy",
            "locationName":  "South Central US",
            "platformVersion":  "19.5",
            "databaseSize":  "@{value=344981504.0; unit=Byte}",
            "ringName":  "Production",
            "appInsightsKey":  ""
        },
```

Figure 7.8 – PowerShell environment list

Some APIs require the use of different methods, such as PUT, POST, or DELETE. Let's look at them with examples. We will learn more about the APIs in the next sections; here, we're just showing examples of PowerShell scripts to provide you with some templates.

Creating or copying an environment

We'll now look at examples of PUT and POST methods being deployed. These methods require you to add a request **body** and ContentType header and are used as follows:

- The PUT method is used to *create* an environment, as described at the following link: https://api.businesscentral.dynamics.com/admin/v2.11/applications/<<Applications Family>>/environments/<<New Environment name>>

- The POST method is used to *copy* an environment, as described at the following link: https://api.businesscentral.dynamics.com/admin/v2.11/applications/<<Applications Family>>/environments/<<Source Environment name>>/copy

The following code snippet shows a PowerShell script to create a new environment with the PUT method:

```
$response = Invoke-WebRequest `
    -Method Put `
    -Uri    https://api.businesscentral.dynamics.com/
admin/v2.11/applications/businesscentral/environments/
MyNewEnvironment `
    -Headers @{Authorization=("Bearer $Token")} `
    -ContentType "application/json" `
    -Body   (@{
            EnvironmentType = "Sandbox"
            CountryCode     = "US"
        } | ConvertTo-Json)
Write-Host (ConvertTo-Json (ConvertFrom-Json $response.
Content))
```

We use the POST method in the same way to copy an environment, as illustrated in the following code snippet:

```
$response = Invoke-WebRequest `
    -Method Post `
    -Uri    https://api.businesscentral.dynamics.com/
admin/v2.11/applications/businesscentral/environments/
MySourceEnvironment/copy `
    -Headers @{Authorization=("Bearer $Token")} `
    -ContentType "application/json" `
    -Body   (@{
            EnvironmentName = "MyNewCopy"
            Type    = "Sandbox"
        } | ConvertTo-Json)
Write-Host (ConvertTo-Json (ConvertFrom-Json $response.
Content))
```

You can check country codes to create necessary localizations here:

```
https://docs.microsoft.com/en-us/dynamics365/business-central/
dev-itpro/compliance/apptest-countries-and-translations
```

Deleting an environment

This section will show you a sample of the DELETE method usage. You can also find out more information at this link: `https://api.businesscentral.dynamics.com/admin/v2.11/applications/<<Applications family>>/environments/<<Environment name>>`.

Using this construction, I could delete my `ProdCopy` sandbox environment with the `BusinessCentral` application family with the next script:

```
$response = Invoke-WebRequest `
    -Method Delete `
    -Uri    https://api.businesscentral.dynamics.com/admin/
v2.11/applications/businesscentral/environments/ProdCopy `
    -Headers @{Authorization=("Bearer $Token")}
Write-Host (ConvertTo-Json (ConvertFrom-Json $response.
Content))
```

This has a scheduled deletion as a result, as shown in the following screenshot:

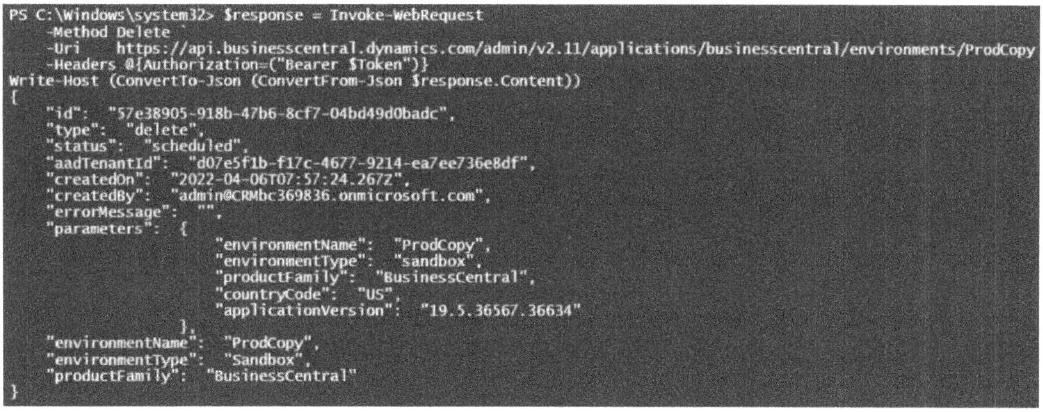

```
PS C:\Windows\system32> $response = Invoke-WebRequest
    -Method Delete
    -Uri    https://api.businesscentral.dynamics.com/admin/v2.11/applications/businesscentral/environments/ProdCopy
    -Headers @{Authorization=("Bearer $Token")}
Write-Host (ConvertTo-Json (ConvertFrom-Json $response.Content))
{
    "id":  "57e38905-918b-47b6-8cf7-04bd49d0badc",
    "type":  "delete",
    "status":  "scheduled",
    "aadTenantId":  "d07e5f1b-f17c-4677-9214-ea7ee736e8df",
    "createdOn":  "2022-04-06T07:57:24.267Z",
    "createdBy":  "admin@CRMbc369836.onmicrosoft.com",
    "errorMessage":  "",
    "parameters":  {
                    "environmentName":  "ProdCopy",
                    "environmentType":  "sandbox",
                    "productFamily":  "BusinessCentral",
                    "countryCode":  "US",
                    "applicationVersion":  "19.5.36567.36634"
                },
    "environmentName":  "ProdCopy",
    "environmentType":  "Sandbox",
    "productFamily":  "BusinessCentral"
}
```

Figure 7.9 – PowerShell delete environment response

You can use the authentication part in your scenarios and change the `Invoke-WebRequest` part with the needed APIs and methods for your routine automation.

Postman

Postman is a tool that you can use to test your APIs. It has a user-friendly interface and you can download it here: `https://www.postman.com/downloads/`.

In this section, we will learn how to connect to the **Admin Center** with **Postman** and call APIs. Follow these next steps:

1. Open Postman, then click on + and create a new request window with the `GET` method and the following **Uniform Resource Locator** (**URL**) that we used to get the environment list previously: `https://api.businesscentral.dynamics.com/admin/v2.11/applications/environments`. You can see an illustration of this in the following screenshot:

Figure 7.10 – Postman request window

2. Click on the **Authorization** tab and choose the **OAuth 2.0** type, as illustrated in the following screenshot:

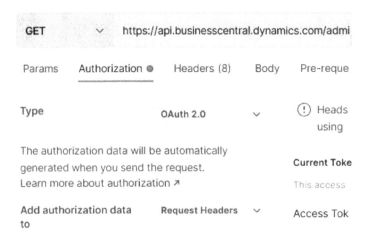

Figure 7.11 – Postman authorization window

3.   On the right side of the **Authorization** tab, fill in the following values:

- **Token Name**: Any name; I just used just a simple `BC`

- **Grant Type**: **Authorization Code**

- **Callback URL**: `BusinessCentralWebServiceClient://auth`

- **Auth URL**: `https://login.windows.net/<<Your tenant ID>>/oauth2/authorize?resource=https://api.businesscentral.dynamics.com`

- **Access Token URL**: `https://login.windows.net/<<Your tenant ID>>/oauth2/token?resource=https://api.businesscentral.dynamics.com`

- **Client ID**: Your app (client) **identifier (ID)**

- **Client Secret**: Leave this blank

- **Scope**: Leave this blank

- **State**: `State`

- **Client Authentication**: **Send client credentials in body**

4.   Then, click on **Get New Access Token**.

The process is illustrated in the following screenshot:

Configuration Options ●   Advanced Options

Token Name   BC

Grant Type   **Authorization Code**   ⌄

Callback URL ⓘ   BusinessCentralWebServiceClient://auth

                              ☐ Authorize using browser

Auth URL ⓘ   https://login.windows.net/d07e5f1b-f17c-4(...

Access Token URL ⓘ   https://login.windows.net/d07e5f1b-f17c-4(...

Client ID ⓘ   810debd2-630d-4c3d-b645-bae2c869... ⚠

Client Secret ⓘ   Client Secret

Scope ⓘ   e.g. read:org

State ⓘ   State

Client Authentication   **Send client credentials in body**   ⌄

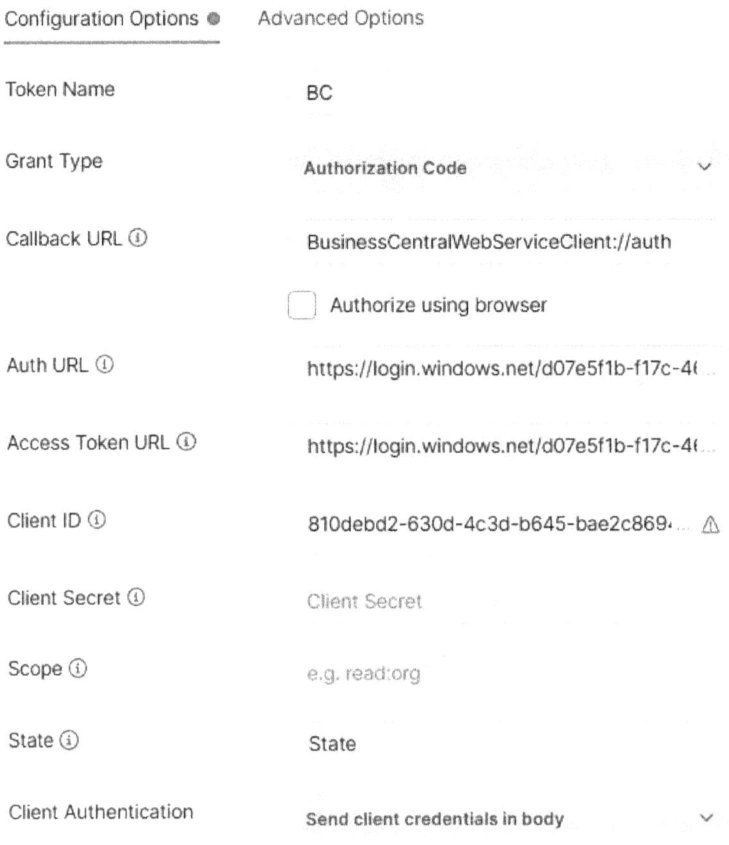

Figure 7.12 – Configuring a new token

5. A sign-in window will open. Input your account and password. After that, a
    token window will appear. Click on the **Use Token** button, as illustrated in the
    following screenshot:

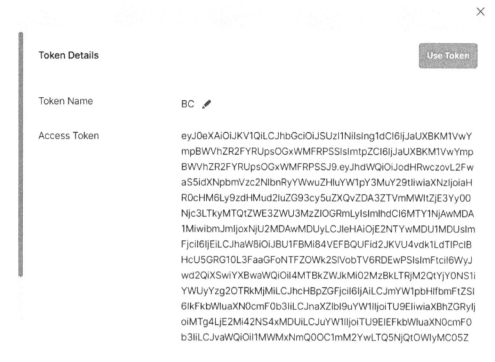

Token Details   Use Token

Token Name   BC ✏

Access Token   eyJ0eXAiOiJKV1QiLCJhbGciOiJSUzI1NiIsIng1dCI6IjJaUXBKM1VwY
mpBWVhZR2FYRUpsOGxWMFRPSSIsImtpZCl6IjJaUXBKM1VwYmp
BWVhZR2FYRUpsOGxWMFRPSSJ9.eyJhdWQiOiJodHRwczovL2Fw
aS5idXNpbmVzc2NlbnRyYWwuZHluYW1pY3MuY29tIiwiaXNzIjoiaH
R0cHM6Ly9zdHMud2luZG93cy5uZXQvZDA3ZTVmMWWItZjE3Yy00
Njc3LTkyMTQtZWE3ZWU3MzZlOGRmLyIsImlhdCI6MTY1NjAwMDA
1MiwibmJmIjoxNjU2MDAwMDUyLCJleHAiOjE2NTYwMDU1MDUsIm
Fjcil6IjEjEiLCJhaW8iOiJBU1FBMi84VEFQQUFid2JJKVU4vdk1LdTIPclB
HcU5GRG10L3FaaGFoNTFFZOWk2SlVobTV6RDEwPSIsImFtcil6WyJ
wd2QiXSwiYXBwaWQiOil4MTBkZWJkMi02MzBkLTRjNjM2QtYjY0NS1i
YWUyYzg2OTRkMjMiLCJhcHBpZGFjcil6IjAiLCJmYW1pbHlfbmFtZSl
6IkFkbWluaXN0cmF0b3liiLCJnaXZlbl9uYW1lljoiTU9EliwiaXBhZGRyIj
oiMTg4LjE2Mi42NS4xMDUiLCJuYW1lljoiTU9EIEFkbWluaXN0cmF0b
3liiLCJvaWQiOil1MWMxNmQ0OOC1mM2YwLTQ5NjQtOWIyMC05Z

Figure 7.13 – Managing access tokens

6. Send a request with the **Send** button. Check the response—it should look like this:

```
Body  Cookies  Headers (11)  Test Results              Status: 200 OK  Time: 1879 ms  Size: 2.45 KB
Pretty  Raw  Preview  Visualize  JSON  ∨
 1  {
 2      "value": [
 3          {
 4              "friendlyName": "CRMbc369836-v20-3",
 5              "type": "Sandbox",
 6              "name": "v20-3",
 7              "countryCode": "US",
 8              "applicationFamily": "BusinessCentral",
 9              "aadTenantId": "d07e5f1b-f17c-4677-9214-ea7ee736e8df",
10              "applicationVersion": "20.1.39764.39901",
11              "status": "Active",
12              "webClientLoginUrl": "https://businesscentral.dynamics.com/d07e5f1b-f17c-4677-9214-ea7ee736e8df/v20-3",
13              "webServiceUrl": "https://api.businesscentral.dynamics.com/v2.0/d07e5f1b-f17c-4677-9214-ea7ee736e8df/v20-3",
14              "locationName": "South Central US",
15              "platformVersion": "20.1",
16              "databaseSize": {
17                  "value": 617611264.0,
18                  "unit": "Byte"
19              },
```

Figure 7.14 – Postman environment list

7. We can use other methods as well—for example, DELETE. Change the method to DELETE.

8. Change the URL to https://api.businesscentral.dynamics. com/admin/v2.11/applications/businesscentral/ environments/<<Your environment name>> and press **Send**, as illustrated in the following screenshot:

Figure 7.15 – Postman DELETE request

You will see a scheduled deletion as a response, as illustrated here:

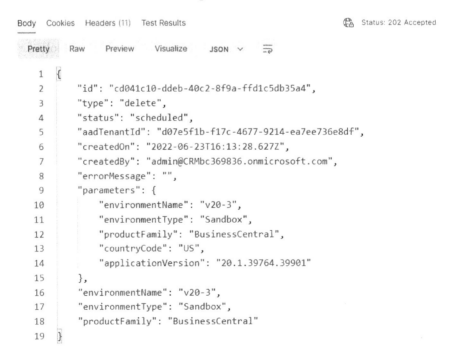

Figure 7.16 – Postman DELETE response

9. We can test PUT and POST method usage by creating and copying environment interfaces. Change the method to PUT and the URL to https://api.businesscentral.dynamics.com/admin/v2.11/applications/businesscentral/environments/<<Your new environment name>>, and on the **Headers** tab, add a Content-Type header of application/json, as illustrated in the following screenshot:

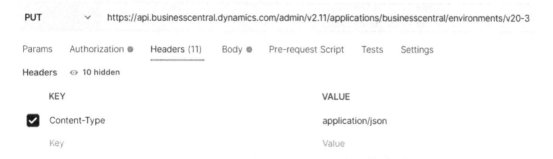

Figure 7.17 – Postman PUT request

10. Click on the **Body** tab. Choose **raw** and fill in the new environment parameters, then click on **Send**. You can see an illustration of this in the following screenshot:

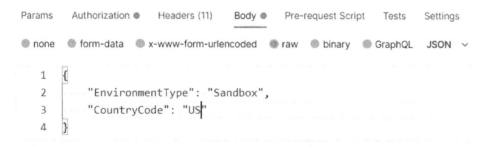

Figure 7.18 – Postman Body section

You will see your scheduled environment has been created, as illustrated in the following screenshot:

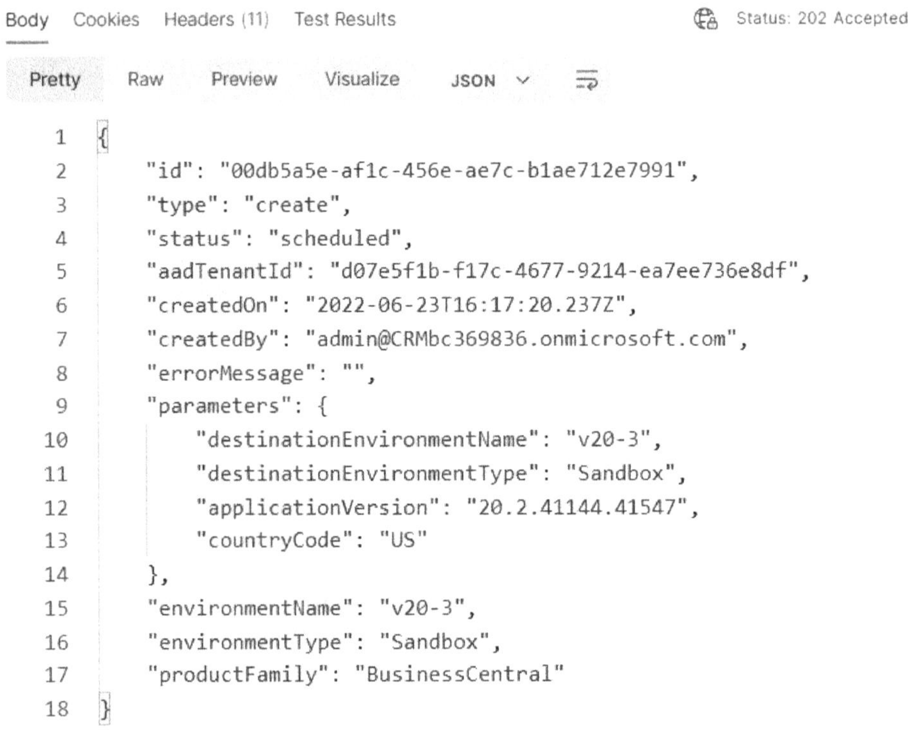
Body
Cookies
Headers (11)
Test Results
Status: 202 Accepted
Pretty
Raw
Preview
Visualize
JSON
{
"id": "00db5a5e-af1c-456e-ae7c-b1ae712e7991",
"type": "create",
"status": "scheduled",
"aadTenantId": "d07e5f1b-f17c-4677-9214-ea7ee736e8df",
"createdOn": "2022-06-23T16:17:20.237Z",
"createdBy": "admin@CRMbc369836.onmicrosoft.com",
"errorMessage": "",
"parameters": {
"destinationEnvironmentName": "v20-3",
"destinationEnvironmentType": "Sandbox",
"applicationVersion": "20.2.41144.41547",
"countryCode": "US"
},
"environmentName": "v20-3",
"environmentType": "Sandbox",
"productFamily": "BusinessCentral"
}

Figure 7.19 – Postman create environment response

In the same way, you can copy the environment using the `POST` method and copying the environment API.

You are now able to call APIs with different methods and from different tools. Next, we will find out more about which APIs the Admin Center provides for us, starting with application APIs.

Application APIs

In this section, we meet the APIs related to **application management** (**AM**)—lists, installation, and other operations. You can always check the latest changes in the APIs— more details are available here: `https://docs.microsoft.com/en-us/dynamics365/business-central/dev-itpro/administration/administration-center-api_app_management`.

Here are some commonly used parameters:

- {applicationFamily}: Application family of the environment. This is usually "businesscentral", but it could be different for **independent software vendor (ISV)** solutions.

- {environmentName}: Name of your environment.

Further, I will give you an API name, a short description (if the name is not obvious), the required method, headers, a URL, and the body's parameters. Let's get started.

Installed apps list

Here are details of the apps list that was installed in your environment:

- Method: GET

- URL: https://api.businesscentral.dynamics.com/admin/ v2.11/applications/{applicationFamily}/environments/ {environmentName}/apps

The result of the execution will be a list of your installed apps with their IDs, which you can use for further actions. You can see an example list here:

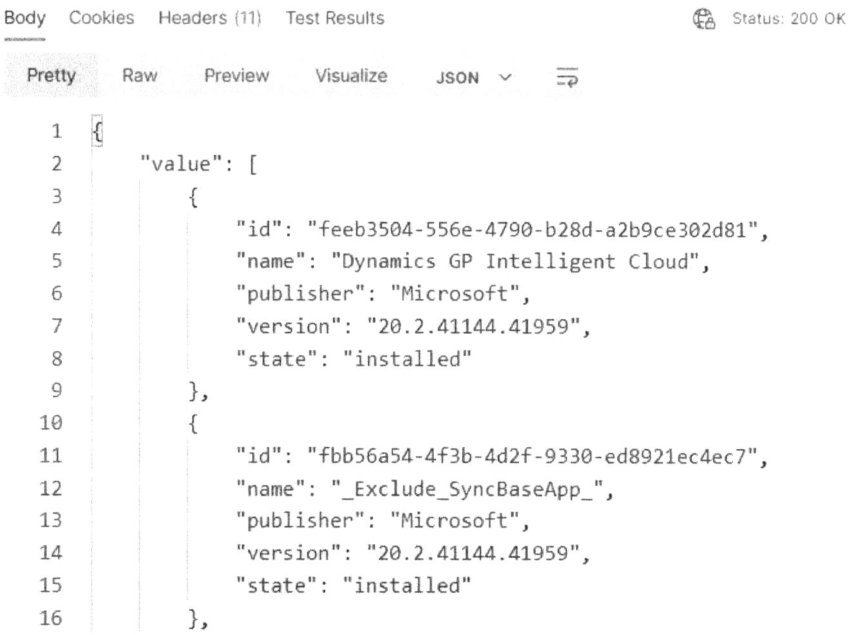

Figure 7.20 – Postman apps list response

Installing an app

This API is used to install an app with a selected app ID to the selected environment. It has the following parameters:

- Headers: `Content-Type: application/json`

- Method: `POST`

- URL: `https://api.businesscentral.dynamics.com/admin/v2.11/applications/{applicationFamily}/environments/{environmentName}/apps/{appId}/install`

- Body parameters, as described in the following table:

Parameter name	Optional or mandatory	Value example	Comment
`targetVersion`	Optional	`1.2.3.4`	Will install the latest version if not provided.
`useEnvironmentUpdateWindow`	Mandatory	`false/true`	If `false`, will be installed immediately; otherwise, will be installed in the environment update window.
`allowPreviewVersion`	Mandatory	`false/true`	If `true`, `targetVersion` becomes mandatory.
`installOrUpdateNeededDependencies`	Mandatory	`false/true`	If `true`, will install or update all the app's dependencies; otherwise, will return a dependencies list in the error message.
`acceptIsvEula`	Mandatory	`false/true`	You must set this as `true` to install the app.
`languageId`	Optional	`en-US`	Specifies local ID language code.

Table 7.1

Uninstalling an app

This API is used to uninstall a selected app from the selected environment. It has the following parameters:

- Headers: Content-Type: application/json

- Method: POST

- URL: https://api.businesscentral.dynamics.com/admin/ v2.11/applications/{applicationFamily}/environments/ {environmentName}/apps/{appId}/uninstall

- Body parameters, as described in the following table:

Parameter name	Optional or mandatory	Value example	Comment
useEnvironmentUpdateWindow	Mandatory	false/ true	If false, will be uninstalled immediately; otherwise, will be installed in the environment update window.
uninstallDependents	Mandatory	false/ true	If true, will uninstall dependent apps.
deleteData	Mandatory	false/ true	Deletes the app data if true.

Table 7.2

Using the app ID from the apps list response, shown in the following screenshot, we can uninstall the app:

Figure 7.21 – Postman uninstall app response

Getting available app updates

This API is used to turn back information about new available app versions. It has the following parameters:

- Method: GET

- URL: https://api.businesscentral.dynamics.com/admin/
  v2.11/applications/{applicationFamily}/environments/
  {environmentName}/apps/availableUpdates

Updating an app

This API is used to update a selected app from the selected environment. It has the following parameters:

- Headers: Content-Type: application/json

- Method: POST

- URL: `https://api.businesscentral.dynamics.com/admin/v2.11/applications/{applicationFamily}/environments/{environmentName}/apps/{appId}/update`

- Body parameters, as described in the following table:

Parameter name	Optional or mandatory	Value example	Comment
`targetVersion`	Optional	`1.2.3.4`	Will install the latest version if not provided.
`useEnvironmentUpdateWindow`	Mandatory	`false`/ `true`	If `false`, will be installed immediately; otherwise, will be installed in the environment update window.
`allowPreviewVersion`	Mandatory	`false`/ `true`	If `true`, `targetVersion` becomes mandatory.
`installOrUpdateNeededDependencies`	Mandatory	`false`/ `true`	If `true`, will install or update all the app's dependencies; otherwise, will return a dependencies list in the error message.

Table 7.3

Getting app operations

This API gives you information about installing, uninstalling, and updating apps. It has the following parameters:

- Method: `GET`

- URL: `https://api.businesscentral.dynamics.com/admin/v2.11/applications/{applicationFamily}/environments/{environmentName}/apps/{appID}/operations/{operationID}`

`OperationID` is not a mandatory parameter. It responds with a single operation if set.

Environment operations APIs

In this section, we meet APIs for environment operations—listing, creating, deleting, backup, and so on.

Detailed information about environment operations updates can be found here:

`https://docs.microsoft.com/en-us/dynamics365/business-central/dev-itpro/administration/administration-center-api_environments`

Environment list

This API is used to return a list of your environments. You could specify the application family to get environments only related to it. It has the following parameters:

- Method: `GET`

- URL without application family: `https://api.businesscentral.dynamics.com/admin/v2.11/applications/environments`

  URL with application family: `https://api.businesscentral.dynamics.com/admin/v2.11/applications/{applicationFamily}/environments`

Getting an environment

This API is used to return environment details selected by the application family and a specified name. It has the following parameters:

- Method: `GET`

- URL: `https://api.businesscentral.dynamics.com/admin/v2.11/applications/{applicationFamily}/environments/{environmentName}`

Creating an environment

This API is used to create a new environment. It has the following parameters:

- Headers: `Content-Type: application/json`

- Method: `PUT`

- URL: `https://api.businesscentral.dynamics.com/admin/v2.11/applications/{applicationFamily}/environments/{environmentName}`

- Body parameters, as described in the following table:

Parameter name	Optional or mandatory	Value example	Comment
`environmentType`	Mandatory	`"Production"; Sandbox"`	Type of environment to create.
`countryCode`	Mandatory	`"US"`	Environment localization.
`ringName`	Optional	`"Preview"`	Environments group. For sandboxes, could be `"Preview"`.
`applicationVersion`	Optional	`1.2.3.4`	Version to create. Will use the latest version if not provided.

Table 7.4

On failure, it returns different error codes explaining why an environment could not be created.

Deleting an environment

This API is used to delete the environment.

It has the following parameters:

- Method: `DELETE`
- URL: `https://api.businesscentral.dynamics.com/admin/v2.11/applications/{applicationFamily}/environments/{environmentName}`

On failure, it returns different error codes explaining why the environment cannot be deleted.

Copying an environment

This API is used to copy an environment from another environment.

It has the following parameters:

- Headers: `Content-Type: application/json`
- Method: `POST`

- URL: `https://api.businesscentral.dynamics.com/admin/v2.11/applications/{applicationFamily}/environments/{sourceEnvironmentName}/copy`

- Body parameters, as described in the following table:

Parameter name	Optional or mandatory	Value example	Comment
type	Mandatory	"Production", "Sandbox"	Type of environment to create
environmentName	Mandatory	"MyCopy"	Name of the created environment

Table 7.5

On failure, it returns different error codes explaining why the environment cannot be copied.

Renaming an environment

This API is used to rename an existing environment.

It has the following parameters:

- Headers: `Content-Type: application/json`

- Method: `POST`

- URL: `https://api.businesscentral.dynamics.com/admin/v2.11/applications/{applicationFamily}/environments/{EnvironmentName}/rename`

- Body parameters, as described in the following table:

Parameter name	Optional or mandatory	Value example	Comment
NewEnvironmentName	Mandatory	"RenamedSandbox"	New name for the environment

Table 7.6

Restoring an environment

This API is used to restore the environment from a restoration point.

It has the following parameters:

- Headers: `Content-Type: application/json`

- Method: `POST`

- URL: `https://api.businesscentral.dynamics.com/admin/v2.11/applications/{applicationFamily}/environments/{EnvironmentName}/restore`

- Body parameters, as described in the following table:

Parameter name	Optional or mandatory	Value example	Comment
`EnvironmentName`	Mandatory	`"RestoredSandbox"`	New name for the environment.
`EnvironmentType`	Mandatory	`"Production"`, `"Sandbox"`	Type of environment to create.
`PointInTime`	Mandatory	`"2022-04-12T20:00:00Z"`	Restore point. Accepts **International Organization for Standardization** (**ISO**) 8601 format with time in **Coordinated Universal Time** (**UTC**).

Table 7.7

Getting available restore periods

This API is used to get the available restore periods for the environment.

It has the following parameters:

- Method: `GET`

- URL: `https://api.businesscentral.dynamics.com/admin/v2.11/applications/{applicationFamily}/environments/{environmentName}/availableRestorePeriods`

Getting used storage for all environments

This API is used to get a list of your environments with used storage in **kilobytes** (**KB**).

It has the following parameters:

- Method: `GET`
- URL: `https://api.businesscentral.dynamics.com/admin/v2.11/environments/usedstorage`

Getting used storage for a selected environment

This API is used to get a single environment, selected by application family and name, with used storage in KB.

It has the following parameters:

- Method: `GET`
- URL: `https://api.businesscentral.dynamics.com/admin/v2.11/applications/{applicationFamily}/environments/{environmentName}/usedstorage`

Getting quotas

This API is used to get tenant quotas: number of environments, and total available storage.

It has the following parameters:

- Method: `GET`
- URL: `https://api.businesscentral.dynamics.com/admin/v2.11/environments/quotas`

Getting environment operations

This API is used to get an environment operations log (copy, create, modify, update, and so on).

It has the following parameters:

- Method: `GET`
- URL: `https://api.businesscentral.dynamics.com/admin/v2.11/applications/{applicationFamily}/environments/{environmentName}/operations`

Getting operations for all environments

This API is used to get all environment operations logs (copy, create, modify, update, and so on).

It has the following parameters:

- Method: `GET`
- URL: `https://api.businesscentral.dynamics.com/admin/v2.11/applications/{applicationFamily}/environments/operations`

Getting environment database export numbers

This API is used to get the number of available database exports with the number used.

It has the following parameters:

- Method: `GET`
- URL: `https://api.businesscentral.dynamics.com/admin/v2.11/exports/applications/{applicationFamily}/environments/{environmentName}/metrics`

Getting environment database export history

This API is used to get who exported a database and when. Parameters are passed with the URL.

It has the following parameters:

- Method: `POST`
- URL: `https://api.businesscentral.dynamics.com/admin/v2.11/exports/history?start={startTime}&end={endTime}`

  `startTime` and `endTime` are `datetime` parameters in UTC.

Creating a database export

This API is used to create your environment database export.

It has the following parameters:

- Headers: `Content-Type: application/json`
- Method: `POST`

- URL: `https://api.businesscentral.dynamics.com/admin/v2.11/exports/applications/{applicationFamily}/environments/{EnvironmentName}`
- Body parameters, as described in the following table:

Parameter name	Optional or mandatory	Value example	Comment
`storageAccountSasUri`	Mandatory		Storage account **shared access signature** (SAS) **Uniform Resource Identifier (URI)**
`container`	Mandatory		Container name to be created
`Blob`	Mandatory	`"*.bacpac"`	Blob name to be created

Table 7.8

Environment settings APIs

This API group is used to get or update different environment settings.

Detailed information on the environment settings API is available here:

`https://docs.microsoft.com/en-us/dynamics365/business-central/dev-itpro/administration/administration-center-api_environment_settings`

Getting an environment update window

This API responds with the environment update preferred start and end time in UTC.

It has the following parameters:

- Method: `GET`
- URL: `https://api.businesscentral.dynamics.com/admin/v2.11/applications/{applicationFamily}/environments/{environmentName}/settings/upgrade`

Changing an environment update window

This API is used to change your environment update window.

It has the following parameters:

- Headers: `Content-Type: application/json`

- Method: `PUT`

- URL: `https://api.businesscentral.dynamics.com/admin/ v2.11/applications/{applicationFamily}/environments/ {EnvironmentName}/settings/upgrade`

- Body parameters, as described in the following table:

Parameter name	Optional or mandatory	Value example	Comment
preferredStartTimeUtc	Mandatory		Start of update window
preferredEndTimeUtc	Mandatory		End of update window

Table 7.9

Here, I set a new update window for my sandbox environment:

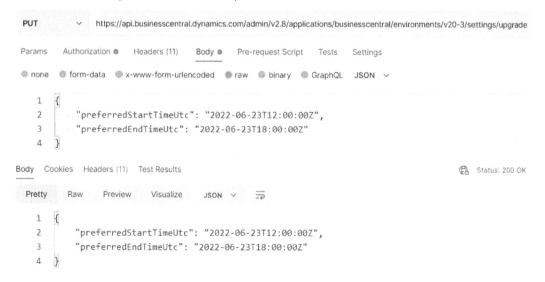

Figure 7.22 – Postman setting a new upgrade window

Changing the environment Application Insights key

This API is used to change the Application Insights key in the selected environment.

It has the following parameters:

- Headers: `Content-Type: application/json`
- Method: `POST`
- URL: `https://api.businesscentral.dynamics.com/admin/v2.11/applications/{applicationFamily}/environments/{EnvironmentName}/settings/appinsightskey`
- Body parameters, as described in the following table:

Parameter name	Optional or mandatory	Value example	Comment
key	Mandatory		Application Insights key

Table 7.10

Getting an environment security group

This API responds with an Azure AD security group, assigned to the environment.

It has the following parameters:

- Method: `GET`
- URL: `https://api.businesscentral.dynamics.com/admin/v2.8/applications/{applicationFamily}/environments/{environmentName}/settings/securitygroupaccess`

Setting an environment security group

This API is used to set a selected Azure AD group as a security group for the environment.

It has the following parameters:

- Headers: `Content-Type: application/json`
- Method: `POST`
- URL: `https://api.businesscentral.dynamics.com/admin/v2.8/applications/{applicationFamily}/environments/{environmentName}/settings/securitygroupaccess`

- A body parameter, as described in the following table:

Parameter name	Optional or mandatory	Value example	Comment
value	Mandatory		**Globally unique ID (GUID)** of Azure AD group

Table 7.11

Deleting an environment security group

This API is used to remove a security group from the environment.

It has the following parameters:

- Method: DELETE

- URL: https://api.businesscentral.dynamics.com/admin/ v2.8/applications/{applicationFamily}/environments/ {environmentName}/settings/securitygroupaccess

Other APIs

This section contains APIs that you can use to perform some other operations.

Information on other APIs can be found here:

https://docs.microsoft.com/en-us/dynamics365/business- central/dev-itpro/administration/administration-center-api_ notifications

Notification settings list

This API is used to get all notification recipients that were assigned to the tenant.

It has the following parameters:

- Method: GET

- URL: https://api.businesscentral.dynamics.com/admin/v2.11/ settings/notification/recipients

Creating a notification recipient

This API is used to add a notification recipient to the tenant.

It has the following parameters:

- Headers: `Content-Type: application/json`

- Method: `PUT`

- URL: `https://api.businesscentral.dynamics.com/admin/v2.11/settings/notification/recipients`

- Body parameters, as described in the following table:

Parameter name	Optional or mandatory	Value example	Comment
`Email`	Mandatory	`John.Doe@ awara-it.com`	Recipient's email
`Name`	Mandatory	`"John Doe"`	Recipient's name

Table 7.12

This is a sample of a notification recipient creation:

Figure 7.23 – Postman creating a recipient

Deleting a notification recipient

This API is used to remove a notification recipient from a tenant.

It has the following parameters:

- Method: `DELETE`
- URL: `https://api.businesscentral.dynamics.com/admin/v2.11/settings/notification/recipients/{RecipientID}`

And this way, you can delete the recipient, as illustrated in the following screenshot:

Figure 7.24 – Postman deleting a recipient

Getting scheduled update information

This API is used to get details about scheduled updates for the selected environment.

It has the following parameters:

- Method: `GET`
- URL: `https://api.businesscentral.dynamics.com/admin/v2.11/applications/{applicationFamily}/environments/{environmentName}/upgrade`

Rescheduling updates

This API is used to change an update date.

It has the following parameters:

- Method: `PUT`

- URL: `https://api.businesscentral.dynamics.com/admin/v2.11/applications/{applicationFamily}/environments/{environmentName}/upgrade`

- Body parameters, as described in the following table:

Parameter name	Optional or mandatory	Value example	Comment
`runOn`	Mandatory	`"2022-04-12T00:00:00Z"`	Sets new upgrade date
`ignoreUpgradeWindow`	Mandatory	`True/False`	Ignores existing upgrade window if `True`

Table 7.13

Getting active sessions

This API is used to get a list of active sessions for the selected environment.

It has the following parameters:

- Method: `GET`

- URL: `https://api.businesscentral.dynamics.com/admin/v2.11/applications/{applicationFamily}/environments/{environmentName}/sessions`

Getting details about a selected session

This API is used to get details about a selected session.

It has the following parameters:

- Method: `GET`

- URL: `https://api.businesscentral.dynamics.com/admin/v2.11/applications/{applicationFamily}/environments/{environmentName}/sessions/{SessionID}`

Killing a session

This API removes an active session selected by session ID.

It has the following parameters:

- Method: `DELETE`
- URL: `https://api.businesscentral.dynamics.com/admin/v2.11/applications/{applicationFamily}/environments/{environmentName}/sessions/{SessionID}`

Getting an environment support contact

This API is used to get the name and email of the environment's support contact.

It has the following parameters:

- Method: `GET`
- URL: `https://api.businesscentral.dynamics.com/admin/v2.11/support/applications/{applicationFamily}/environments/{environmentName}/supportcontact`

Setting an environment support contact

This API is used to set the name and email of the environment's support contact.

It has the following parameters:

- Method: `PUT`
- URL: `https://api.businesscentral.dynamics.com/admin/v2.11/support/applications/{applicationFamily}/environments/{environmentName}/supportcontact`
- Body parameters, as described in the following table:

Parameter name	Optional or mandatory	Value example	Comment
`name`	Mandatory	John Doe	Sets support contact name
`email`	Mandatory	`John.Doe@awara-it.com`	Sets support contact email
`url`	Mandatory	`http://www.awara-it.com`	Sets additional support URL

Table 7.14

Getting environment telemetry

This API is used to get basic environment telemetry.

It has the following parameters:

- Method: `GET`

- URL: `https://api.businesscentral.dynamics.com/ admin/v2.11/applications/{applicationFamily}/ environments/{environmentName}/ telemetry?startDateUtc={startdate}&endDateUtc={enddate} &logCategory={category}`

  Here, `{startdate}` is at the beginning of the selection window, `{enddate}` is at the end of the selection window (both parameters are of the datetime type), and `{category}` has an `"All"` value.

Summary

This is the last chapter of the first part of the book. You made a big step forward and got a new professional skill—you learned how to automate your administration routine. You are able to create automation scripts and perform single calls with **Postman**. You know which tasks you are able to automate. And after all this information about the **Admin Center**, we will move on to cloud migration in the second part of the book.

Part 2: Dynamics 365 Business Central Cloud Migration Tool

The cloud migration tool allows you to migrate data from on-premises to a SaaS environment. In this part, you will learn how to set up cloud migration from an on-premises environment to SaaS, run the migration, upgrade data, and fix something that goes wrong.

This part contains the following chapters:

- *Chapter 8, Cloud Migration Schema and Limitations*
- *Chapter 9, Cloud Migration Setup*
- *Chapter 10, Migration Process*
- *Chapter 11, The Real Migration Experience*

8
Cloud Migration Schema and Limitations

We start *Part 2* with **Cloud Migration**. Cloud Migration is a special tool that helps you to migrate your data from any on-premises Business Central environment (still supported in version 14) or Dynamics GP to the SaaS environment. It is free of charge and already included in your cloud environment even if you do not need it. You do not need to search for any distributives; the only thing you need to install additionally is a **self-hosted integration runtime** – part of the Cloud Migration tool. You will get a link to download it while you run the setup process.

The main official purpose of the Cloud Migration tool is to help you migrate from your on-premises environment to the cloud, but later we will look at a non-official usage – data consolidation. For some reasons (which we will find out about later), it is not officially recommended, but it works with some exceptions.

You do not need developer skills to use it. Not a single line of code is required; it is pure administration. From the **Business Central 2022 release wave 1**, delegated admins can perform the migration. If you are a Partner, you do not need a separate license for this action anymore.

Cloud Migration made a long journey from quite a raw instrument (when it was called **Intelligent Edge**) to a real business tool that will help you to take your ERP system to the next level. I have been working with the tool since it first appeared, firstly, just to get to know the new technology. Then, me and my team got a chance to use it in a real project. We took the risk and it was worth it.

In this chapter, we are going to cover the following main topics:

- Cloud Migration and how to use it

- Migration schema and the main nodes

- Migration limitations

By the end of this chapter, you will know what Cloud Migration is, along with its parts, migration stages, and limitations.

Technical requirements

To create a migration, you must have the following:

- A Dynamics 365 Business Central SaaS tenant (also works on a demo tenant)

- A **SUPER** user in the Business Central SaaS or a delegated admin with granted consent

- A Microsoft 365 tenant admin account

The currently supported products for migration to SaaS are the following:

- All versions of Dynamics 365 Business Central on-premises.

- Dynamics GP 2015 and newer.

- Dynamics NAV customers are also able to migrate, but they have to upgrade their systems to Dynamics 365 Business Central first. To find out how to do it, you can read the docs page here: `https://docs.microsoft.com/en-us/dynamics365/business-central/dev-itpro/upgrade/upgrading-to-business-central`.

- On-premises SQL Server database:

  - SQL Server 2016 and newer

  - Compatibility level 130 (the compatibility level defines how SQL Server uses certain features)

  - SQL Server authorization

Cloud Migration and how to use it

The Cloud Migration tool lives in your cloud environment. You can find it by searching for the **Cloud Migration Management** page. Before the setup, it should look empty.

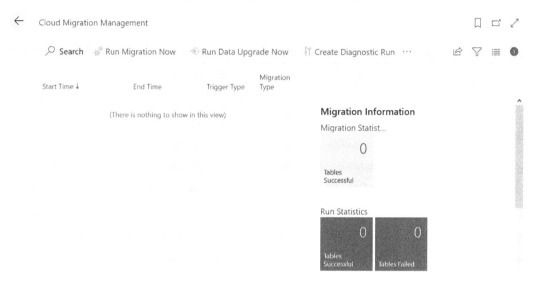

Figure 8.1 – Cloud Migration Management

We can have a look at the backend and open the **Installed Extensions** page.

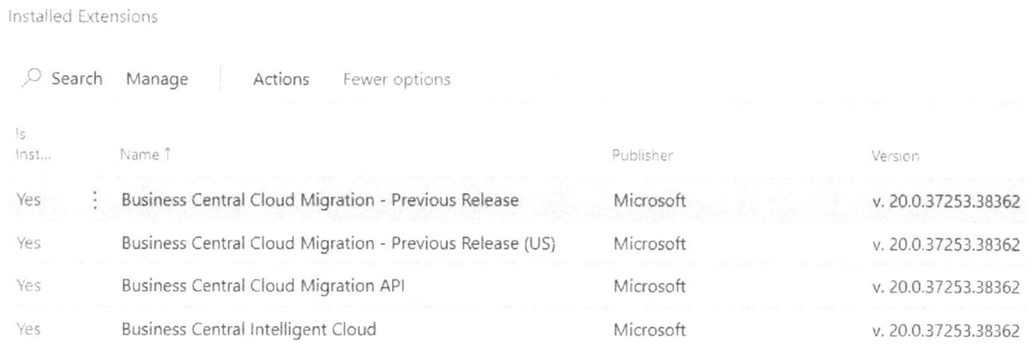

Figure 8.2 – Cloud Migration extensions

You can see three to four extensions here, related to Cloud Migration:

- **Business Central Intelligent Cloud**: The main app.

- **Business Central Cloud Migration - Previous Release**: This app lets you migrate old Business Central on-premises versions to the cloud.

- **Business Central Cloud Migration - Previous Release (Localization Code)**: Not mandatory. You can not use it in every localization. It contains localization specialties for the migration process.

- **Business Central Cloud Migration API**: A new feature, introduced in the **Business Central 2022 release wave 2**. This lets you automate the migration process.

If you migrate from **Dynamics GP**, you need the following extensions:

- **Dynamics GP Intelligent Cloud**: This app is used to migrate data from Dynamics GP.

| Yes | Dynamics GP Intelligent Cloud | Microsoft | v. 20.0.37253.38362 | Global |

Figure 8.3 – Dynamics GP Intelligent Cloud

- **Dynamics GP History SmartLists**: This extension is used to load several predefined SmartLists that will show Microsoft Dynamics SL historical transactions data.

If some of these apps (not just GP) are not installed in your environment, you can find them in the **Extensions Marketplace**.

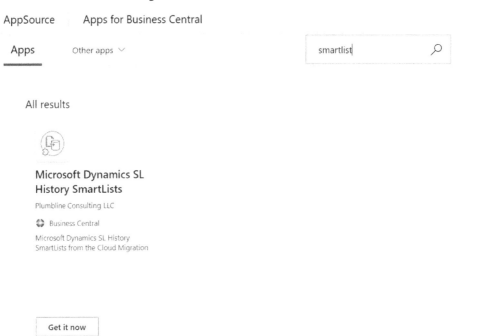

Figure 8.4 – Microsoft Dynamics SL History SmartLists

When running Cloud Migration for the first time, you can do so from the **Assisted Setup** page.

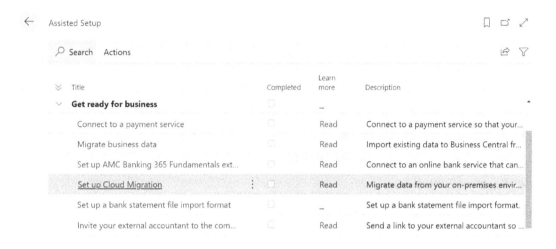

Figure 8.5 – Assisted Setup

This option is just used to run the setup process or to show with a **Completed** mark that setup has been done. The other way is to search for `cloud migration` in the **Tell me what you want to do** search box.

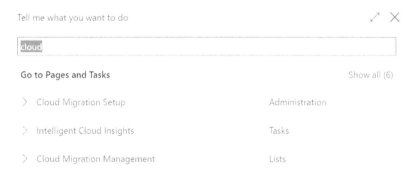

Figure 8.6 – Searching for Cloud Migration

As you see, it has three related pages:

1. **Cloud Migration Setup**: Runs the same page as we have in the **Assisted Setup**
2. **Cloud Migration Management**: Runs the page to handle the migration process
3. **Delegated admin consent for cloud migration**: A special page where the Business Central license owner can grant consent to the **delegated admin** to run Cloud Migration

In this chapter, we will get to know the parts of Cloud Migration that will be commonly used. The usage and details will be described in the next chapters.

Migration schema and the main nodes

The Cloud Migration tool consists of the following nodes:

- On-premises SQL Server database or Azure SQL Database
- Self-hosted integration runtime or Azure runtime
- Azure Data Factory pipelines
- Azure Integration Runtime for Data Factory
- Business Central SaaS

The common migration schema looks like this:

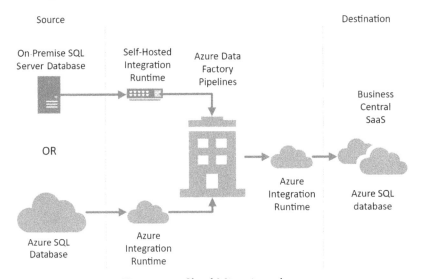

Figure 8.7 – Cloud Migration schema

If you migrate from an on-premises SQL Server database, you must install the **self-hosted integration runtime** and register it with a special authentication key, which you get during the setup, to let Azure Data Factory connect to the database.

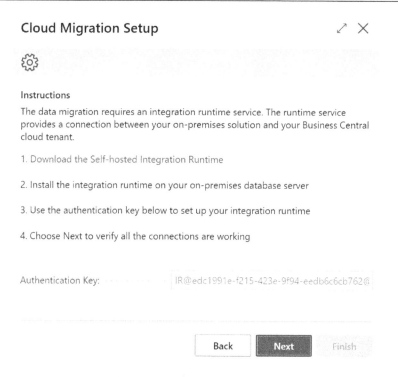

Figure 8.8 – Self-hosted integration runtime authentication key

For Azure SQL Database, the integration runtime is created automatically. Azure Data Factory only uses your SQL Server permissions with the user account that you use for the migration.

Migration limitation

Previously, the main limitation in Cloud Migration was the database size – **80 GB**. Now, you can migrate larger-sized databases; but *can* does not mean *should*.

If you exceed the storage limit, the migration process will not stop; you will have to buy additional capacity with a storage add-on.

Microsoft recommends migrating no more than **30 GB** per migration. But how do you achieve this if your database is large?

1. First of all, you can migrate your companies one by one. Select one company on the initial setup and you will be able to change the companies list to migrate later after the first company migrates.

2. Secondly, delete or archive unneeded data. Sounds easy, but this is a very tough process. Please refer to *Chapter 6*, for the details.

3. You can add the `ReplicateData` = `False` property to your custom tables if you do not want to replicate them.

```
1    table 50117 "AWR_Sales Invoice Line clone"
2    {
3        Caption = 'Sales Invoice Line';
4        Permissions = TableData "Item Ledger Entry" = r,
5                      TableData "Value Entry" = r;
6        ReplicateData = false;
7
```

Figure 8.9 – ReplicateData table property

If you want to migrate a large database and cannot shrink it, you may ask a support team to assist you in this process.

You will probably be wondering about custom tables and field migration. Data is migrated if it exists in both environments – on-premises and SaaS. If you have custom tables and fields in an on-premises environment and want to migrate them to the cloud, you must deploy the same changes to the SaaS first. If your on-premises Business Central is an April 2019 release, you must redevelop your C/AL customizations to AL if you already developed them before the data migration.

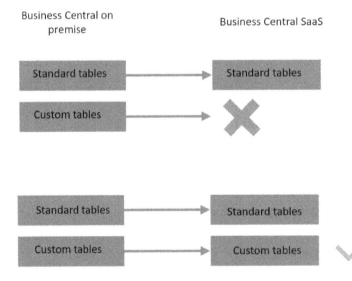

Figure 8.10 – Cloud Migration schema

Some releases ago, the Cloud Migration tool allowed regularly scheduled runs. This meant that you were able to have a regularly updating copy of your on-premises database in the cloud without any lines of code. Now, this functionality had been cut out because if migration runs while the environment is upgrading, this could cause serious issues. For managed and automated migration runs, APIs could be used.

> **Note**
>
> Migration isn't limited between regions from on-premises to SaaS, for example, RU on-premises to UK SaaS. Data will migrate but only for equal schema elements.

You could even collect data from different databases into one SaaS environment, for example, to consolidate data for some analysis. Remember that per-database tables will be overwritten each time and some links could be corrupted.

> **Note**
>
> Per-database table data applies to all companies in the database. Per-company table data applies only to the current company.

> **Additional Information**
>
> Microsoft gives extra details about Cloud Migration in Microsoft Docs: `https://docs.microsoft.com/en-us/dynamics365/business-central/dev-itpro/administration/migrate-data`.
>
> For Microsoft Partners, the product team is available on Yammer. You will be surprised by how responsive they are there: `https://aka.ms/bccloudmigrationyammer`.

Summary

Now you know what Cloud migration is and what nodes it consists of. It is no longer a black box for you. You've seen that Cloud Migration is a very good tool but has some limitations that you need to be aware of. In the next chapter, you will learn, in detail, how to set up the migration process.

9
Cloud Migration Setup

Now you have learned what cloud migration is from the previous chapter, we can start to set it up. You do not need developer skills here, as everything performs with an assisted setup. Here, we will learn important things such as connection string construction and how to troubleshoot typical issues that could occur during the setup process. In addition, we will investigate how you can automate the setup process with cloud migration **application programming interfaces** (**APIs**).

In this chapter, we are going to cover the following main topics:

- Prerequisite 1—SQL Server setup

- Prerequisite 2—Connection string construction

- Prerequisite 3—Delegated admin consent

- Migration setup and the **self-hosted integration runtime** (**SHIR**)

- Cloud migration APIs

By the end of this chapter, you will know how to set up cloud migration to move your data from an on-premises environment to the cloud. We'll start by setting up the prerequisites first.

Prerequisite 1 – SQL Server setup

Before a cloud migration setup can run, we must check our **on-premise SQL Server** settings. If you use **Azure SQL Server**, then skip this section.

Our check includes three points, as outlined here:

1. **SQL Server version**

    You must check that you have SQL Server 2016 or higher since older versions are not supported. You can upgrade your SQL Server version if needed.

    To check your SQL Server version, use Command Prompt, as follows:

    - Type `SQLCMD -S servername\instancename`

    - Type `select @@version`

    - Type `go`

    The process is illustrated in the following screenshot:

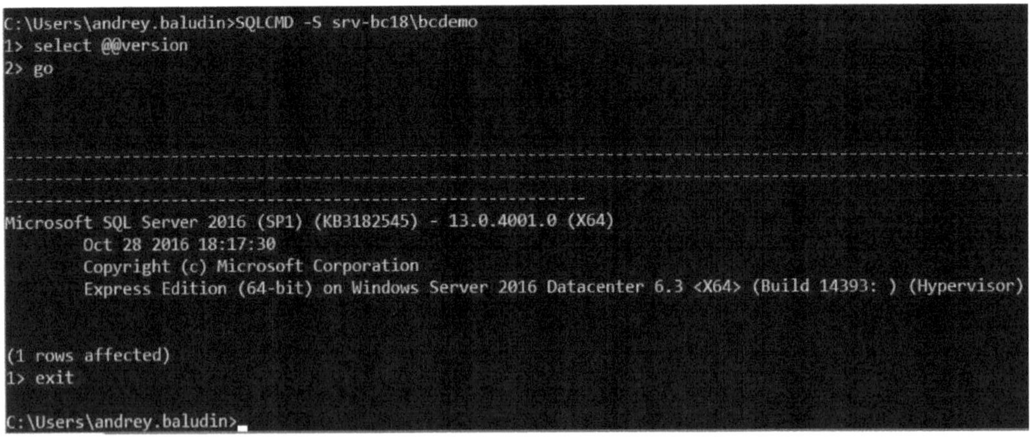

Figure 9.1 – SQL Server version check

2. **Database compatibility level**

    Run **SQL Server Management Studio** (**SSMS**), click on your database, and choose **Properties**. Then, choose **Options** and check the **Compatibility level** value. This must be **SQL Server 2016 (130)** or higher. Change it if needed and apply the changes.

The process is illustrated in the following screenshot:

Figure 9.2 – Database compatibility level

3. **SQL Server authorization**

    Open your server **Properties** section and choose the **Security** tab. Check that you have a **Server authentication** value of **SQL Server and Windows Authentication mode**. If not, change the value and restart SQL Server.

    The process is illustrated in the following screenshot:

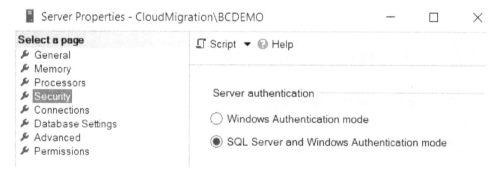

Figure 9.3 – SQL Server authorization

Then, we need a user with **SQL Server authentication** to run the migration. You can use some existing ones or create a new one. Here's how you can do this:

1. Open the **Security** folder of SQL Server.

2. Choose **Logins**, and after right-clicking, choose **New Login**.

3. Choose **SQL Server authentication**, input a new password, and don't forget to uncheck the **User must change password at next login** box!

The process is illustrated in the following screenshot:

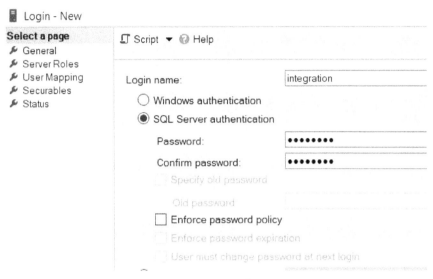

Figure 9.4 – New login

4. On the **User Mapping** tab, map the database that you want to migrate. Then, click **OK** to create a new login.

The process is illustrated in the following screenshot:

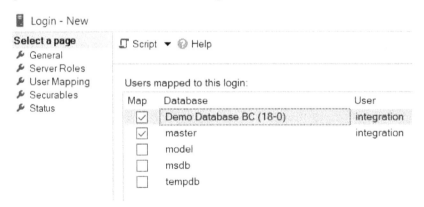

Figure 9.5 – User mapping

With this, we have checked the critical points and can construct a connection string to our database.

Prerequisite 2 – Connection string construction

You will need a connection string when you run a cloud migration setup, so, we need to prepare this beforehand.

A connection string consists of four parameters, as outlined here:

- **Server name**
- **Database name**
- **Username**
- **User password**

The full view looks like this:

```
server={Server name\Instance name};Initial Catalog ={Database
name};user id ={User name};password ={User password};
```

As an example, I have SQL Server `CloudMigration` with a `BCDEMO` instance. My Business Central database is named `Demo Database BC (18-0)`. To migrate this database, I created an `Integration` user with the password `integration123`. Using these parameters, my connection string will look like this:

```
server=CloudMigration\BCDEMO;Initial Catalog ="Demo Database BC
(18-0)";user id =integration;password =integration123;
```

If I have my database in **Azure SQL**, the connection string might look like this:

```
Server=tcp:CloudMigration.database.windows.
net,1433;Database="Demo Database BC (18-0)";User ID=
integration@CloudMigration;Password= integration123;
```

Prerequisite 3 – Delegated admin consent

If you run a cloud migration setup with a delegated admin account, you need to get consent from a user with a Business Central license and with a SUPER permissions set. This user should open the **Delegated admin consent for cloud migration** page and click **Grant Consent**. Consent could be revoked with a **Revoke Consent** action.

The following screenshot illustrates the process:

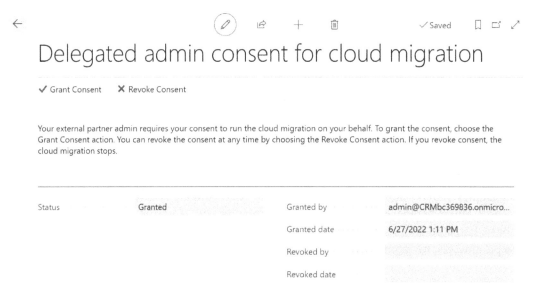

Figure 9.6 – Delegated admin consent

We now have all the prerequisites and can start the migration setup.

Migration setup and the SHIR

In this section, we will learn how to start the migration setup and the steps it requires. I will divide the setup process into the following three steps:

1. Running the cloud migration wizard
2. Installing the SHIR
3. Finalizing the cloud migration setup

Let's learn about each step in detail.

Running the cloud migration wizard

To run the cloud migration wizard, proceed as follows:

1. Open your target environment Dynamics 365 Business Central **software as a service (SaaS)** and go to the **Assisted Setup** page. Choose the **Set up Cloud Migration** option from the **Get ready for business** section, as illustrated in the next screenshot:

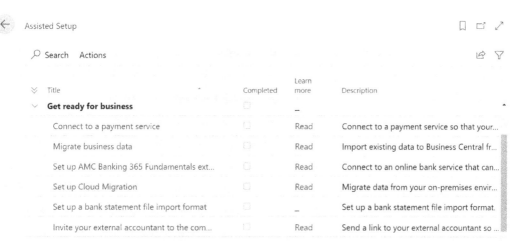

Figure 9.7 – Assisted setup

Important Note

You need to perform this setup in the Business Central cloud environment, and not on-premises.

---

2. As a first step, you must choose a product that you want to migrate data from. Click on the ellipsis (**...**) near the **Product** value and choose a data source, as illustrated in the following screenshot:

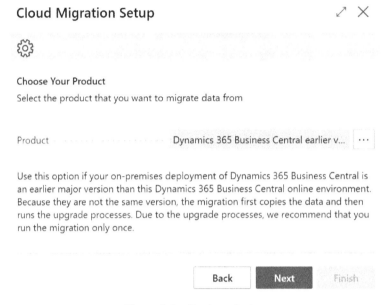

Figure 9.8 – Product choice

You can migrate from the following:

- **Dynamics 365 Business Central version**, if your on-premise version is the same as the cloud version

- **Dynamics 365 Business Central earlier versions** (v14 and newer)

- **Dynamics Great Plains** (**Dynamics GP**)

The following screenshot illustrates these choices:

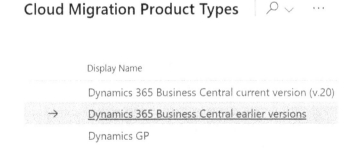

Figure 9.9 – Product types

3. Now, we need to define a database connection. Here's how we go about this:

  - Choose an **SQL Configuration** value—**SQL Server** or **Azure SQL**.

  - Paste the **SQL Connection String** value that we created in the previous section.

  - Leave **Integration Runtime Name** blank if you're running the setup for the first time, or input this if you've already created this.

  - Click **Next**.

The process is illustrated in the following screenshot:

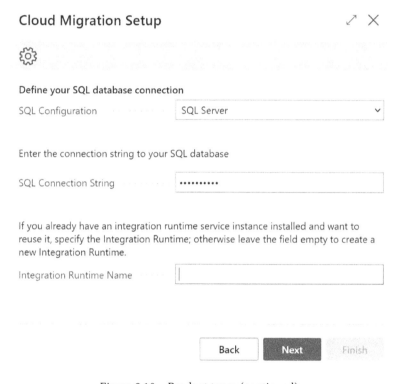

Figure 9.10 – Product types (continued)

4. If you left **Integration Runtime Name** blank in the previous step and the database source is **SQL Server**, you will see a window with a download link for the **SHIR** (see https://www.microsoft.com/en-us/download/details.aspx?id=39717 for more details).

5. Next, you can copy the **Authentication Key** value somewhere.

The process is illustrated in the following screenshot:

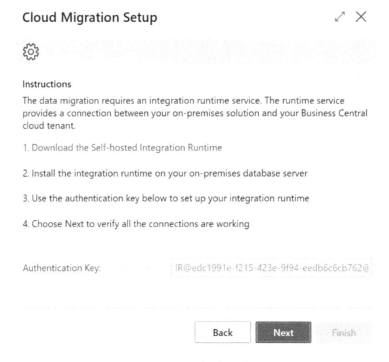

Figure 9.11 – Downloading the SHIR

Installing the SHIR

1. Now we need to install the SHIR on the server, where your version of SQL Server is installed. Click on the link from point **1** in *Figure 9.11*, or use this **Uniform Resource Locator** (**URL**): https://www.microsoft.com/en-us/ download/details.aspx?id=39717.

Download the SHIR and run the installation. The process is illustrated in the following screenshot:

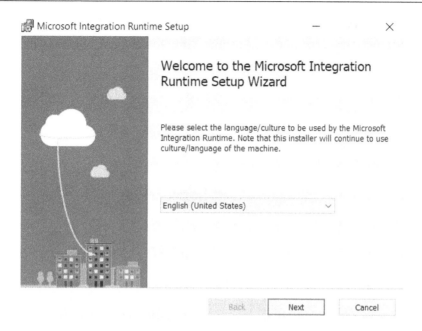

Figure 9.12 – Installing the SHIR

2. When the installation ends, you will need to register your runtime. Paste your **authentication key** from **Cloud Migration Setup** and click **Register**.

The process is illustrated in the following screenshot:

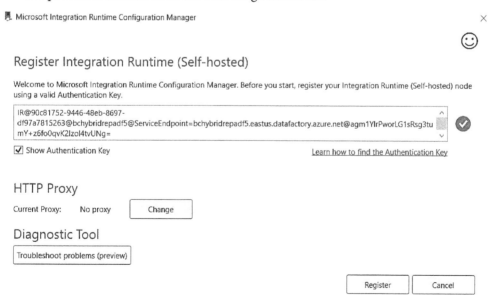

Figure 9.13 – Registering the SHIR

3. You will see your runtime name (something such as `msft1a6720t02101126`) and node name. You can copy the runtime name somewhere, but you will be able to make it to the further screen.

4. Click **Finish**.

The process is illustrated in the following screenshot:

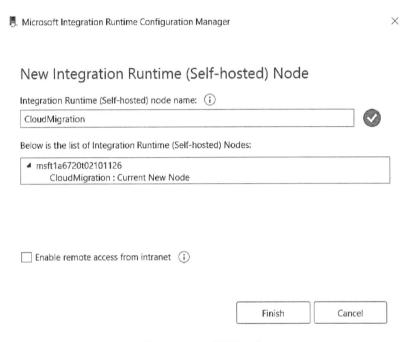

Figure 9.14 – SHIR node

5. You should see a message that your SHIR was successfully registered. Click **Launch Configuration Manager**.

The process is illustrated in the following screenshot:

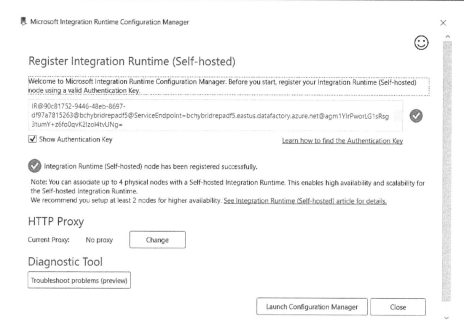

Figure 9.15 – Finishing SHIR registration

6. You can use **Configuration Manager** to get the SHIR name if you need to run the setup again. In addition, it shows the current SHIR state. Here, if you see a green mark, as in the following screenshot, then everything is OK:

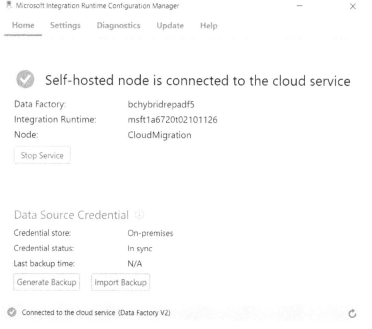

Figure 9.16 – SHIR Configuration Manager

7. If you have any concerns about your connection string, you can test it on the **Diagnostics** tab.

8. Now, choose **SqlServer** for the connection, and input your **SQL Server name** value and **database name** instance.

9. Next, choose basic authentication, input your username and password, and click **Test**. If a green mark appears near the **Test** button, as in the next screenshot, then your parameters are correct:

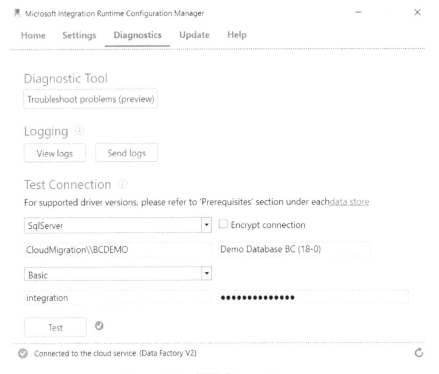

Figure 9.17 – SHIR Diagnostics page

> **Important Note**
> Use a double slash between the SQL Server name and instance, like this:
> `CloudMigration\\BCDEMO`.

Finalizing the cloud migration setup

To finalize the cloud migration setup, proceed as follows:

1. Turn back to the **Cloud Migration setup**. Input your **Integration Runtime Name** value if needed and click **Next**. You need to wait while it gets the data.

    The process is illustrated in the following screenshot:

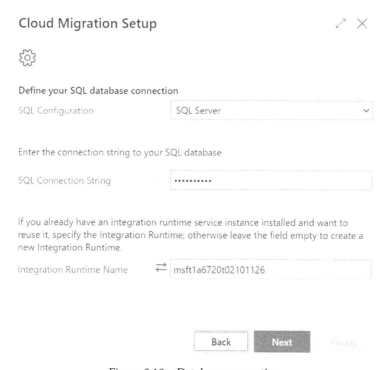

Figure 9.18 – Database connection

2. On a successful run, you will see a list of companies. Choose the required companies or click **Migrate all companies** and click **Next**, as illustrated in the following screenshot:

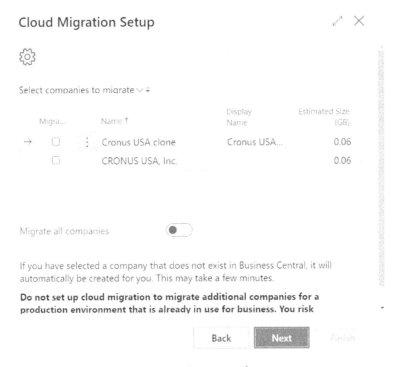

Figure 9.19 – Company selection

Important Note

In this step, sometimes I got a connection error even with the correct connection string. If you get some strange error, then just try to run the connection again.

3. After you click **Next**, you will get a message about the successful setup. Click **Finish**. The process is illustrated in the following screenshot:

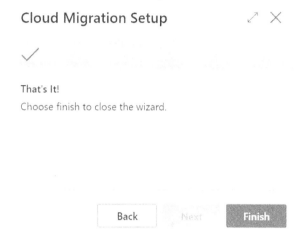

Figure 9.20 – Finishing the migration setup

4. You can open the **Companies** list and ensure that the companies you have selected before are appearing on the list. You can see in the following screenshot that the companies appear on the list:

Figure 9.21 – Companies list

Remember that data is not migrated yet. We will continue with data migration in the next chapter. Microsoft describes the migration process with the next scheme:

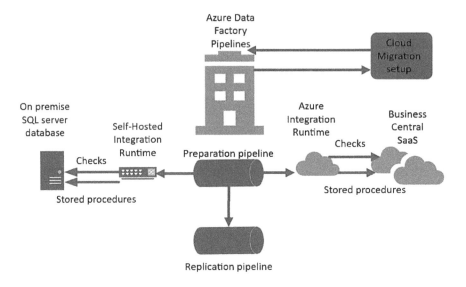

Figure 9.22 – Cloud migration setup schema

Cloud migration APIs

If you need to run the setup for multiple customers, databases, or environments, you can use APIs to automate your work. They have been implemented since **Business Central 2022 wave 1**. In *Chapter 7*, you learned how to call APIs in different ways, so I just describe which APIs are available and which parameters they have. You can read additional details about cloud migration APIs in *Microsoft Docs* at https://docs.microsoft.com/en-us/dynamics365/business-central/dev-itpro/administration/cloudmigrationapi/cloud-migration-api-overview.

In the next chapter, you will also find a PowerShell script that uses all these APIs.

Cloud migration setup API

As we need to perform several actions to complete the setup, this API also consists of three steps, as detailed next.

First step

Here, we input basic parameters—source migration product, connection string, SQL Server type, and the company where we want to run the setup, as follows:

- Headers: `Content-Type:; application/json`
- Method: `POST`
- URL: `https://api.businesscentral.dynamics.com/v2.0/ {TenantID}/{EnvironmentName}/api/microsoft/cloudMigration/ v1.0/companies({CompanyID})/setupCloudMigration`

The following parameters are also required:

- `{TenantID}`—Your Business Central SaaS tenant **identifier (ID)**
- `{EnvironmentName}`—Target environment name
- `{CompanyID}`—ID of the cloud company from where you are running the cloud migration

The body parameters are listed in the following table:

Parameter name	Optional or mandatory	Value example	Comment
`productId`	Mandatory	`"DynamicsBCLast"`, `"DynamicsBC"`, `"DynamicsGP"`	Type of the environment to migrate from
`sqlServerType`	Mandatory	`"SQLServer"`, `"AzureSQL"`	SQL Server type
`sqlConnectionString`	Mandatory		Your connection string

Table 9.1

After you send the request, you will get extra parameters in the response, as illustrated here:

```
{
    "id":"{SetupRecordId}",
    "productId":"{ProductID}",
    "sqlServerType":"{SqlServerType}",
    "sqlConnectionString":"{SqlConnectionString}"
    "runtimeName":"{RuntimeName}",
    "runtimeKey":"{RuntimeKey}"
}
```

Second step

In the second step, we provide the actions needed to set up the SHIR. You can skip this step if you use Azure SQL because it uses an Azure integration runtime.

The parameters are listed here:

- Headers:
- `Content-Type`: `application/json`
- `If-match`: `Etag`
- Method: `PATCH`
- URL: `https://api.businesscentral.dynamics.com/v2.0/{TenantID}/{EnvironmentName}/api/microsoft/cloudMigration/v1.0/companies({CompanyID})/setupCloudMigration({SetupRecordID})`

The following parameter is also required:

- `{SetupRecordID}`—Setup record ID, which you got in the first step

The body parameters are listed in the following table:

Parameter name	Optional or mandatory	Value example	Comment
`productId`	Mandatory	`"DynamicsBCLast"`, `"DynamicsBC"`, `"DynamicsGP"`	Type of environment to migrate from
`sqlServerType`	Mandatory	`"SQLServer"`, `"AzureSQL"`	SQL Server type
`sqlConnectionString`	Mandatory		Your connection string
`runtimeName`	Mandatory	`"msft1a6720t02101126"`	Your SHIR name

Table 9.2

Third step

With the third step, we finalize the setup.

The parameters are listed here:

- Method: `POST`
- URL: `https://api.businesscentral.dynamics.com/ v2.0/{TenantID}/{EnvironmentName}/api/microsoft/ cloudMigration/v1.0/companies({CompanyID})/ setupCloudMigration({SetupRecordID})/Microsoft.NAV. completeSetup`

On-premise companies list

You need this API to get information about companies that you could include in the migration.

The parameters are listed here:

- Method: `GET`
- URL: `https://api.businesscentral.dynamics.com/v2.0/ {TenantID}/{EnvironmentName}/api/microsoft/cloudMigration/ v1.0/companies({CompanyID})/cloudMigrationCompanies`

The response is shown in the following code snippet:

```
{
  "id": "{OnPremCompanyId}",
  "name": "{CompanyName}",
  "replicate": false,
  "displayName": "{DislpayName}",
  "estimatedSize":{SizeOfComapny},
  "status": "",
  "created": false
}
```

Including the on-premise company in the cloud migration

After we complete the setup, it is time to choose which companies we want to migrate. This is a two-step API, which includes an on-premises company in the cloud migration process. Run it as many times as you have different companies to migrate.

First step

In the first step, we mark an on-premises company to replicate or remove the replication mark, to exclude the company from the replication process.

The parameters are listed next.

- Here are the headers:
- `Content-Type:` `application/json`
- `If-match:` `Etag`
- Method: `PATCH`
- URL: `https://api.businesscentral.dynamics.com/` `v2.0/{TenantID}/{EnvironmentName}/api/microsoft/` `cloudMigration/v1.0/companies({CompanyID})/` `cloudMigrationCompanies({OnPremCompanyID})`

The following parameter is also required:

- `{OnPremCompanyID}`—ID of the company from the on-premises companies list API response.

The body parameters are listed in the following table:

Parameter name	Optional or mandatory	Value example	Comment
`Replicate`	Mandatory	`true/false`	`true` if replication is needed. `false` to exclude the company from the replication

Table 9.3

Second step

Finalize the company selection in a similar way to how we finalized the setup in the previous API.

Here are the headers:

- Method: `POST`

- URL: `https://api.businesscentral.dynamics.com/v2.0/{TenantID}/{EnvironmentName}/api/microsoft/cloudMigration/v1.0/companies({CompanyID})/cloudMigrationCompanies({OnPremCompanyID})/Microsoft.NAV.createCompaniesMarkedForReplication`

Summary

You know how to set up **cloud migration** and everything related to it. You are now able to deploy the **SHIR**, construct a connection string, and test it. Finally, you know how to automate this work with **cloud migration APIs**.

In the next chapter, you will learn how to migrate your data after you have done the setup, and I will present you with a very useful script from the Microsoft team that will help you with the routine automation.

10
Migration Process

In the previous chapter, we completed the cloud migration setup. This action does not mean that your data is already in the cloud even if you see your on-premise company on the list. Here, we will demystify the migration process and how your data moves from on-premise to the cloud and upgrade the process because the table schema might be different in a newer environment. Also, you will learn how to automate migration runs.

In this chapter, we are going to cover the following main topics:

- Data migration
- Data upgrade
- Typical fails and how to resolve them
- Automating data migration

By the end of this chapter, you will understand how to set up cloud migration so that you can move your data from an on-premise environment to the cloud.

Data migration

The full migration process is divided into two stages:

- Data migration
- Data upgrade

You can run data migration several times, for example, by synchronizing your companies from on-premise to SaaS one by one, by cleaning your data in the destination point, and by only upgrading your data at the end as a final step, when you are sure that everything has been correctly migrated.

Microsoft describes the migration process with the following schema:

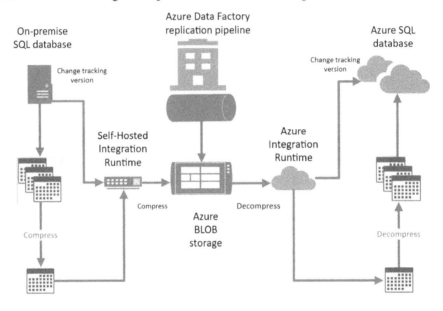

Figure 10.1 – The data migration process

You might have a question about the per database tables synchronization. Take a look at *Figure 10.2*. It is from *Microsoft Business Central Launch Event 2022 wave 1*. What if we sync the Company 1 tables with the per database tables and then sync the Company 2 tables with the per database tables again? If we overwrite the per database tables each time, data could have been already changed in such tables and links could have become broken. The answer is that all tables that synchronize in one moment per database table do not have double replication, so your data will migrate correctly:

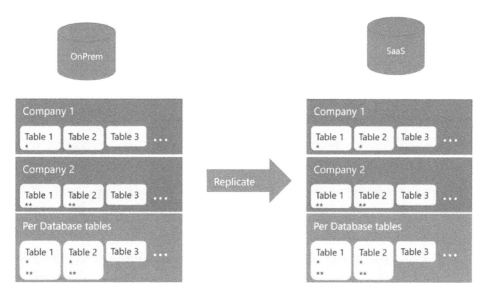

Figure 10.2 – Table replication

To start the migration process, open the **Cloud Migration Management** page in the **Business Central SaaS**. Here, you can either **Run Migration Now** or **Create Diagnostic Run** to do more data verification before the normal migration run, to decrease the risk of a failed migration.

Important Note

Before you start the data migration, open the **Admin Portal**, and if you have a pending target environment update, postpone it. Similarly, if you need to upgrade your cloud environment, disable the cloud migration from the **Cloud Migration Management** page.

You can view the **Cloud Migration Management** page here:

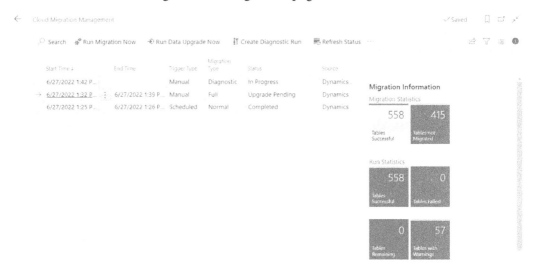

Figure 10.3 – The Cloud Migration management page

In addition, this page contains other useful actions. Let's look at the most important ones:

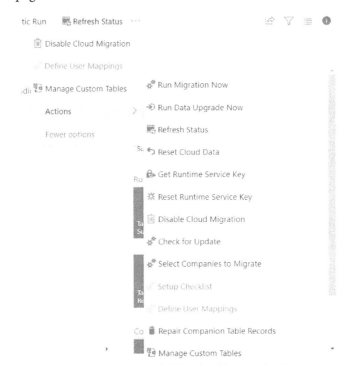

Figure 10.4 – Cloud migration actions

Here is a list of the useful actions:

- **Run Data Upgrade Now**: This starts the data upgrade.
- **Refresh Status**: You can use this action to refresh the state of your running migration operations.
- **Reset Cloud Data**: This resets the migration-enabled data in the cloud tenant.
- **Get Runtime Service Key**: If you forgot to copy the Integration Runtime authorization key, you can find it here.
- **Reset Runtime Service Key**: This resets the Integration Runtime authorization key.
- **Disable Cloud Migration**: Use this action to disable the cloud migration process; for example, before an environment upgrade.
- **Select Companies to Migrate**: Here, you can change your companies to migrate choice.
- **Manage Custom Tables**: Use this action to manage the custom table's mappings:

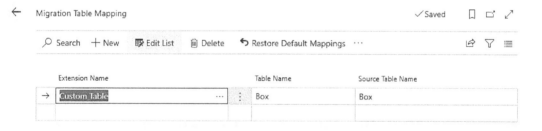

Figure 10.5 – Migration table mappings

Do you remember the **ReplicateData** property for the custom tables? By default, it is **True**. So, all your custom tables will try to migrate. You can exclude them from the replication process by setting this property to **False**. I say *try to migrate* because you must have the same tables in the SaaS environment. You can map them with the **Manage Custom Tables** functionality if they have different names (for example, if you use Business Central 14 and moved your C/AL code to AL before migration).

To start the migration process, select **Run Migration Now**. Answer **Yes**, and wait until you see a confirmation message:

Figure 10.6 – A migration confirmation message

You will see that the migration is in progress:

Figure 10.7 – Cloud migration in progress

The migration time depends on the size of your database. It could be anywhere from several minutes for a **Cronus Demo database** to several hours for a real database. Click on **Refresh Status** to update the status information.

After the migration, the status will change to **Data Repair Pending**:

Figure 10.8 – Cloud migration data repair

And, a bit later, it will change to **Upgrade Pending**:

Figure 10.9 – Cloud migration upgrade pending

You can check the migration details in the **Migration Information** fact box:

Figure 10.10 – Migration Information

The numbers are informative; click on them to see the details:

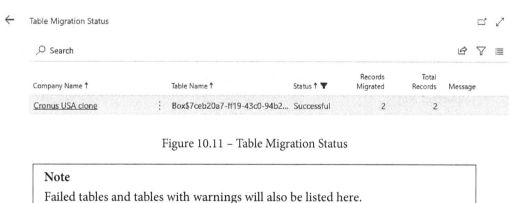

Figure 10.11 – Table Migration Status

> **Note**
> Failed tables and tables with warnings will also be listed here.

Before the data upgrades, you can verify or clean up the migrated data. It is highly recommended that you create a copy of the production environment as a sandbox and run the data upgrade there first.

You can migrate data several times to add deltas or extra companies and then run the data upgrade at the end.

Data upgrade

We can illustrate the data upgrade process with the following schema:

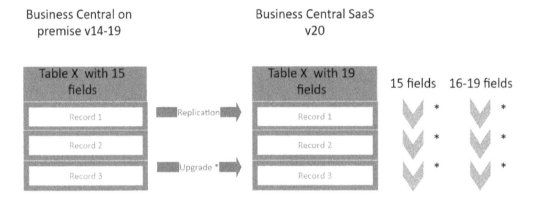

Figure 10.12 – The data upgrade schema

We have table X with **15** fields in the on-premises Business Central instance. It should be an old version from 14 to 19. We migrate data to the new Business Central SaaS v.20 where table X already has **19** fields because new functionality was added. We migrate 15 fields from the on-premises instance. After that, we upgrade data to fill the last 16–19 fields in the new environment.

> **Important Note**
>
> Before you start the data upgrade, open the **Admin Portal**, and if you have a pending target environment update, postpone it. Similarly, if you need to upgrade your cloud environment, disable the cloud migration from the **Cloud Migration Management** page.

After you migrate your data, open the **Admin Portal** and create a copy of your target migration environment. Test the data upgrade in this copy first. This is not a mandatory step, but it is highly recommended. You can restore this copy at any point if something went wrong.

To start the data upgrade process, click on the **Run Data Upgrade Now** action on the **Cloud Migration management** page. You must confirm that you have migrated all the companies and are ready to finish this process:

> ⑦ The upgrade must be triggered as the last step, because you'll not be able to migrate further data after the upgrade. Before you start the upgrade, make sure that you have moved all companies that you want to move.
>
> Are you sure that you want to proceed?
>
> [Yes]   [No]

Figure 10.13 – Data upgrade confirmation

After the confirmation, your migration status will turn into **Upgrade in Progress**:

Figure 10.14 – Data upgrade status

And after that, the status will change to **Complete**.

If you try to click **Run Migration** again after the data upgrade, you will get an error. You must delete the migrated company (you can do this from the **Companies** list) and start the migration again:

Figure 10.15 – Data migration error

After the migration, all users in the migrated company (except the admin) will be assigned to the **Intelligent Cloud** user group. This means that the database for them becomes read-only. This is made on purpose, to prevent any unexpected data modifications before you complete the migration. After you upgrade the data, you must open the **Users** card and assign normal permission sets to allow them to work as usual.

The last step will be to ask users to stop working in the on-premise environment and provide them with a new connection URL.

Typical failures and how to resolve them

For now, there are not so many problems with cloud migration. I have Microsoft statistics about the most popular issues. They are listed as follows:

- A failed connection from the Self-Hosted Integration runtime to an on-premise database.

  Solution: Test your Integration Runtime connection using the diagnostics tool. It will likely be network issues.

- Self-Hosted Integration runtime setup errors.

  Solution: As in the first point, test your connection. You likely set the wrong authorization key. Check it one more time or generate a new one.

- Wrong connection string.

  Solution: Check all the parameters one more time. It is likely that you mistyped somewhere.

- The database compatibility level is lower than **130**.

  Solution: Check the database compatibility level. Change it, if needed.

- The on-premise database has been replaced after setup.

  Solution: Set up the cloud migration process again.

I also met some strange and unexpected errors during the migration, which were caused by major environment updates. If you are a Partner, you can always ask the development team about such cases in Yammer at `https://aka.ms/bccloudmigrationyammer`.

Automating data migration

As part of the cloud migration setup, Microsoft provides APIs to run data migration and data upgrades. They can be useful if you are planning a bulk migration or migrations for several customers. If you need more details, you can find them in the official Microsoft documentation at `https://docs.microsoft.com/en-us/dynamics365/business-central/dev-itpro/administration/cloudmigrationapi/cloud-migration-api-overview`.

However, here, we will skip their boring description and have more practice.

To make your life easier, Microsoft created a **PowerShell** script with a set of functions for cloud migration. It contains all that you need – the setup, migration, and data upgrade. You can download it from `https://aka.ms/BCCloudMigrationAPI`.

The link contains two files:

- **CloudMigrationStatusText.psm1**: This contains state-of-art graphic design functions.

- **CloudMigrationScripts.ps1**: This contains the main cloud migration functions:

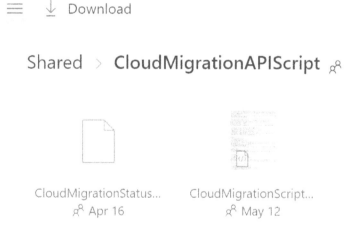

Figure 10.16 – The cloud migration API script

To start using the script, please download both files into one folder to the machine where you have installed or are planning to install the **Integration Runtime**, run **Windows PowerShell ISE** as an administrator, and open **CloudMigrationScripts.ps1**.

You must fill the following parameters with your values:

- **AADTenantID**: This is your Business Central SaaS tenant ID.

- **CurrentUserName**: This is your admin username.

- **CurrentPassword**: This is your admin user password.

- **EnvironmentName**: This is the name of the cloud environment where you plan to migrate your on-premise companies.

- **MainCompanyID**: This is the ID of the cloud company from which you want to run the cloud migration setup.

- **ClientID**: This is the client ID of the app registration, which you use for OAuth 2.0 authorization:

Figure 10.17 – The script parameters

Before the first run, I needed two extra things to make this script work:

1. Add and execute the `Import-Module "MSAL.PS"` line. Microsoft writes about this step in the third line of the script; only in the first run.

2. Check that your app registration in the **Azure Portal**, under the **Manifest** tab, has `allowPublicClient = true`. Otherwise, change it and save your app registration:

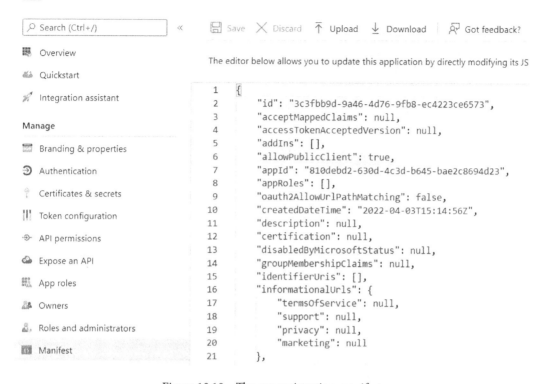

Figure 10.18 – The app registration manifest

After that, run your script with the *F5* key to load the variables. The script contains many functions, but we will start with the common `Run-CloudMigrationE2E` function and a `-SqlConnectionString` parameter. You can revisit how to construct a connection string in the previous chapter. Here is my example:

```
Run-CloudMigrationE2E -SqlConnectionString
"server=CloudMigration\BCDEMO;Initial Catalog ='Demo Database
BC (18-0)';user id =integration;password =YOURPASSWORD;"
```

Next, you must sign in so that the script will run. If you have already done the cloud migration setup, it will skip this step.

As a result, you will see the whole migration process step by step (remember when I told you about state-of-the-art graphic design?):

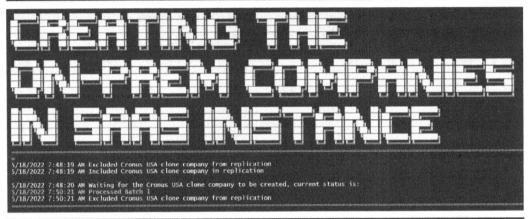

Figure 10.19 – Running the cloud migration script

After the script execution, open your SaaS environment and ensure the company has been migrated. Please investigate the script, and you can create your own scenarios based on the functions it has.

Summary

Now you know what data upgrade is and why you need it. You know the typical issues and are ready to fight them. Finally, you know how to automate your cloud migration with a pre-prepared **PowerShell** script. You are ready to migrate from on-premise **Dynamics 365 Business Central** to the cloud. In the final chapter, I will tell you about my experience with cloud migration.

11
The Real Migration Experience

This is the final chapter of the book. Let's put the technical details, methods, and setups aside and just talk about the real usage of the Cloud Migration tool. I'm proud that I was one of the first users of this tool to implement it in a real project. I want to tell you my story. I hope that you will find it interesting and it will inspire you to not be afraid of new technologies, even if they seem quite raw.

In this chapter, we are going to cover the following main topics:

- Cloud Migration – early beginnings and evolution
- Issues in the course of implementation

Cloud Migration – early beginnings and evolution

The first time I heard about Cloud Migration was after the **Business Central 2018 October** release. It was referred to by the name **Intelligent Edge**. As I understood from different sources, the tool allowed you to replicate your data from NAV 2018, Business Central on-premises, and Dynamics GP to Business Central SaaS, but the table schema strictly had to be the same. Migration failed if the tables had even one different field. This was not very interesting to me, because each project I worked on had some customizations, which made such migrations impossible. And so I forgot about it (for several months at least).

> **Reference**
>
> You can see the old presentation here: `https://docs.microsoft.com/en-us/business-applications-release-notes/october18/dynamics365-business-central/dynamics-intelligent-edge`.

Before the **Business Central 2019 April** release was presented, Microsoft said that Intelligent Edge had been upgraded and the strict schema requirement was no longer the case. This was a big step forward, but without a real usage scenario, it could have ended up being nothing more than a demo for me.

In April 2019, just after the new Business Central release, I came into contact with a new customer. They had headquarters in Europe where they used Business Central SaaS and a branch office in Russia where they used the Business Central 2018 October release on-premises. The task was: *we need a copy of all main tables that are on-premises in the cloud environment, because we want to consolidate data there.*

Which tables were the *main* ones? How often would we need to replicate data? And how many APIs would I need for that? Stunning. This task looked like a big project, until I remembered Intelligent Edge.

Our solution was based on a non-standard tool usage scenario for two reasons:

- We had planned to use scheduled replication each day instead of one-time migration.
- We had planned to migrate to different localizations, which was not officially supported.

We offered this tool to the customer with the following pros and cons:

- **Pros**: Implementation time would be very fast and costs would be much lower than migrating table by table.

- **Cons**: It is a black box. There is a lack of information about real usage and the tool is quite new. There could be some issues during the implementation.

So, we decided to run a demo environment and test it. Intelligent Edge had been renamed **Intelligent Cloud**. We installed the Business Central 2018 October release on-premises with Russian localization and replicated it without errors to the Business Central SaaS 2019 April release with Great Britain localization. This was amazing because the tables' schemas were different.

This is a screenshot from the old setup. As you can see, it has a common version of Business Central and a separate Dynamics NAV 2018 product:

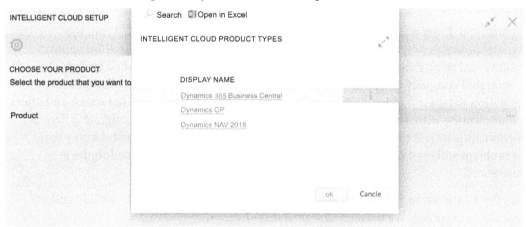

Figure 11.1 – Intelligent Cloud setup

Around that time, I wrote my first blog post about Intelligent Cloud setup. It seems to hold up, despite all the changes in the Cloud Migration tool. If you are interested, you can find it here: `https://community.dynamics.com/business/b/andreysnavblog/posts/intelligent-cloud-setup-for-business-central-spring-release`.

After the successful demo, we presented the solution to the customer. At this stage, we were interested to see how different localizations could be replicated. However, this was no miracle tool, and all tables and fields specialized for Russian localization had been missed. This was expected and the customer said that these tables were not the *main* ones and other concerns were about per-database tables, which could be overwritten, but here we did not meet any issues. In addition, we had a possibility to migrate data to the sandbox always, because it was planned to use it only for analysis as a read-only company.

We decided to accept all the risks and start a project with Intelligent Cloud.

Issues in the course of implementation

I created a copy of the customer's production database and started replication.

First issue

The first issue I met was that on-premises databases were in **Azure SQL**. From the migration admin's point of view, it looked perfect. You don't need to install Integration Runtime and setup runs faster. We were happy until the migration finished and we opened the migrated company in the cloud. The source database had a Cyrillic collation and the destination database had Latin. All Cyrillic text transformed into ??? characters, and after communicating with different people, I realized that there is no way to migrate different collations this way. We had to migrate to the on-premises SQL Server first. I know that this problem still exists because I've heard the same problem raised by colleagues at conferences.

After the migration to SQL Server on-premises, replication finished successfully and no more issues occurred. We decided to start our go-live.

Second issue

The second issue we met was after our SaaS environment got the **2019 October release** update. The customer wrote that replication stopped and wouldn't run again. I decided to run the setup again. In fact, this is a good first-aid tip when something suddenly goes wrong with the replication process.

A surprise was waiting for me when I tried to select a product to migrate from. There were no earlier Dynamics 365 Business Central versions, just the current one. I tried to find the app in the application market but did not succeed. It supported only NAV 2018 and the current Business Central version. When I tried to use the current version, I got this message:

Business Central on-premises must be at least version 15 to use
the cloud migration functionality.

 OK

Figure 11.2 – Migration warning

I raised a support ticket and got an answer that earlier versions would be back after some fixes. In 1 week, we were able to migrate again. After that, I was very careful with major updates. This situation also repeated with **2020 release wave 1**, but I postponed the SaaS upgrade until Intelligent Cloud was upgraded (which happened 2 weeks later).

Third issue

After some minor updates, we ran into a third issue – scheduled replication stopped just after the update. Microsoft explained that an upgrade with working replication could damage data and replication must be stopped before the upgrade. This rule works for now. If you schedule an environment upgrade and do not disable Cloud Migration, the upgrade will fail and re-schedule.

This did not look like a serious issue in the beginning, because running the migration setup when you have everything installed is quite fast. However, to reduce the amount of migrating companies before the upgrade, Microsoft removed the **Scheduled run** feature from Cloud Migration. This was a game-changer for us. I wrote a ticket but got the answer that it is disabled almost forever.

These are screenshots from the past when it was possible to create scheduled migration runs:

Figure 11.3 – Setting up a migration schedule

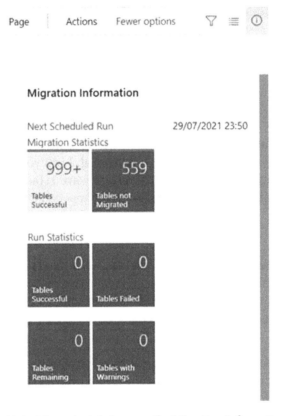

Figure 11.4 – Next scheduled run on the Migration Information tab

5. Enable & Scheduling Migration

The final page in the wizard allows you to enable the migration process and create a schedule for when the data migration should occur. These settings are also available within your Business Central tenant on the **Cloud Migration Management** page. You have the option to schedule migrations daily or weekly.

♀ **Tip**

We recommend that you schedule your data migration for off-peak business hours since it can take many hours to run, depending on the amount of data.

We also recommend that you make sure that all users are logged out of both the source company and the target company.

Figure 11.5 – Microsoft Docs tip about scheduled migration

After a discussion with the customer, we decided to run the migration manually, before they needed fresh numbers in the cloud. The database was not very big and the migration time was about 15 minutes, so it was not a problem.

Since **2022 release wave 1**, it is possible to run the migration through an API call, so you can create scheduled runs again by yourself.

Extra issue

After hearing the experience of my colleagues, I want to share with you one interesting issue. When you create records with Excel import, such as customers or vendors, in an on-premises environment, you can get non-printing characters in the code key fields. They will migrate to the SaaS and you can get some issues when using the `Record.Get()` function there.

To prevent this situation, run the following command in the Business Central administration shell:

```
$result = Invoke-NAVSanitizeField -ServerInstance
YOURSERVERINSTANCE -Tenant YOURTENANT
```

This command removes all non-printing characters from all fields of the code type in the tenant database and returns a list of modified records and tables.

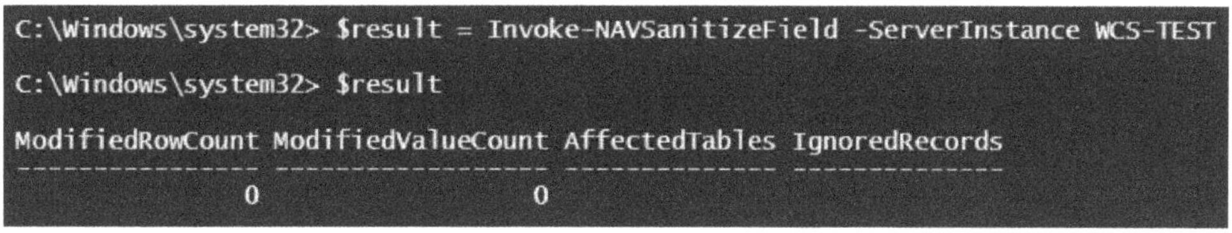

Figure 11.6 – SanitizeField command

You can find more details here: https://demiliani.com/2021/12/14/dynamics-365-business-central-are-you-doing-cloud-data-migration-please-sanitize-your-database-before-doing-it.

Summary

To conclude, I can say that the project was definitely worth it. We dived deep into a new technology, which worked as expected, despite issues that occurred during the exploitation. The customer got exactly what they wanted and saved money. We got a huge amount of experience in using Business Central without writing a single line of code.

I hope that this story inspired you to use non-standard ways of task resolution. Business Central developers must not be limited to using only Business Central. They must know how to use as many Microsoft products as they can. **Azure**, **Dataverse**, **Cloud Migration**, **Power BI** – they all help you to complete some tasks much faster than you would be able to by writing a hundred lines of AL code.

In this book, I tried to collect everything that a Business Central specialist needs for the administration or migration of an environment to the cloud. I will be happy if someday I see this book on somebody's table. Thanks for having me! Good luck, colleagues, and I hope that this book is useful to you.

Index

Other Books You May Enjoy

If you enjoyed this book, you may be interested in these other books by Packt:

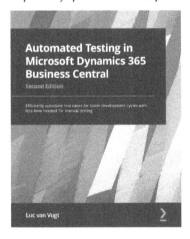

Automated Testing in Microsoft Dynamics 365 Business Central - Second Edition

Luc van Vugt

ISBN: 9781801816427

- Understand the why and when of automated testing
- Discover how test-driven development can help to improve automated testing
- Explore the six pillars of the Testability Framework of Business Central
- Design and write automated tests for Business Central
- Make use of standard automated tests and their helper libraries
- Understand the challenges in testing features that interact with the external world
- Integrate automated tests into your development practice

Microsoft Dynamics 365 Business Central Cookbook

Michael Glue

ISBN: 9781789958546

- Build and deploy Business Central applications
- Use the cloud or local sandbox for application development
- Customize and extend your base Business Central application
- Create external applications that connect to Business Central
- Create automated tests and debug your applications
- Connect to external web services from Business Central

Packt is searching for authors like you

If you're interested in becoming an author for Packt, please visit authors. packtpub.com and apply today. We have worked with thousands of developers and tech professionals, just like you, to help them share their insight with the global tech community. You can make a general application, apply for a specific hot topic that we are recruiting an author for, or submit your own idea.

Share Your Thoughts

Now you've finished *Administrating Microsoft Dynamics 365 Business Central Online*, we'd love to hear your thoughts! Scan the QR code below to go straight to the Amazon review page for this book and share your feedback or leave a review on the site that you purchased it from.

https://packt.link/r/1803234806

Your review is important to us and the tech community and will help us make sure we're delivering excellent quality content.

www.ingramcontent.com/pod-product-compliance
Lightning Source LLC
Chambersburg PA
CBHW060548060326
40690CB00017B/3646